Programming Workflow Applications with Domino

Daniel Giblin and Richard Lam

CRC Press
Taylor & Francis Group
Boca Raton London New York

CRC Press is an imprint of the
Taylor & Francis Group, an **informa** business

CRC Press
Taylor & Francis Group
6000 Broken Sound Parkway NW, Suite 300
Boca Raton, FL 33487-2742

First issued in hardback 2017

Copyright © 2000 by Taylor & Francis
CRC Press is an imprint of Taylor & Francis Group, an Informa business

No claim to original U.S. Government works

ISBN 13: 978-1-138-41250-7 (hbk)
ISBN 13: 978-1-929629-06-0 (pbk)

**Visit the Taylor & Francis Web site at
http://www.taylorandfrancis.com**

**and the CRC Press Web site at
http://www.crcpress.com**

Cover Art Design: Robert Ward

To Chris & Claire.
DTG

For my father.
RBL

Acknowledgments

We would like to thank Peter Fairweather for his encouragement, support, and mentorship. Special thanks to Lei Kuang and Keith Grueneberg for their exceptional Notes knowledge, and to Sherman Alpert, Cindy Conway, and Dan Oblinger for sharing their Java experiences, and to Diane Melley for her guidance and support. We would also like to thank Marsha Haug for her valued contributions. Last, but not least, we recognize those people who sacrificed the most during the preparation of this book: Chris and Kathy.

Preface

Today's business climate is evolving at a rapid pace. The Internet has created new channels for advertising and commerce. It has spawned new paradigms for pricing and delivering products and services. Customers have access to unprecedented amounts of information, and their content expectation levels continue to rise. The result is that businesses must frequently reevaluate their strategies, effect rapid changes in their business processes, and provide timely knowledge and support to an increasingly savvy customer base.

To address these problems, many corporations and organizations have turned to the Web for answers. Connectivity among businesses, suppliers, and customers provides unique opportunities for cooperation and integration. Unfortunately, much web development concentrates on disseminating information through static web pages. Thus, the application of the web for e-commerce, e-support, and e-workflow has been limited. However, the concept of groupware has emerged as a powerful tool for reshaping business processes and customer interactions.

Lotus Notes/Domino™ is the industry leader in groupware technology. Notes Designer, in conjunction with Notes/Domino server, allows development of web *applications*, not simply web pages. These applications not only provide access to static web pages of information, they also handle dynamic content, including email, calendaring, document sharing, etc. When the workflow features for submission, routing, approval, and review of Notes forms and documents are combined with the built-in access control functionality of Notes databases, new models of web-based workflow applications become possible. Kathy Lam, senior IT strategy consultant for IBM Global Services, uses the term "webflow" to describe this class of applications.

This book provides an introduction to workflow application development using the Notes/Domino environment. Following an introduction to the concepts of workflow, we discuss Notes features important to workflow applications. This leads to the design and implementation of a framework for a general workflow management system that can be reused in both existing and new Domino applications. A case study follows the steps involved in designing a workflow application, where the use of the workflow management system is shown in action.

The book is accessible to all Notes developers with at least some practical Notes design experience. A working knowledge of LotusScript or Java is also essential to understand the Notes agent examples on which the workflow engine components are based. We hope the discussions and examples in this book better enable you to harness the power of Notes to create your own sophisticated workflow applications.

This book serves two purposes. First, we describe the basic concepts of workflow as it relates to business processes. We discuss the features and benefits of Lotus Notes as an integrated development environment particularly suited for groupware and workflow application development. We also define a notification scheme for use in simple scenarios in which a full-fledged workflow process would be inappropriate.

Second, we present a complete workflow management system developed within the Notes framework. The framework can be modified, extended, and reused in a variety of workflow applications. We supply and annotate the complete source code for the framework, and we illustrate its use through an example case study.

After reading this book, you should understand the concepts necessary for implementing workflow applications to handle various business processes. Also, you should be able to author your own significant Notes and web-based workflow applications using this book's notification and workflow management schemes.

Table of Contents

Chapter 1

Introduction

You encounter workflow processes every day. Any time you fill out a form and submit it, you initiate a workflow process. Your form must be routed to the appropriate destination. It must be reviewed, signed, forwarded, returned, approved, or rejected. In business, such processes are ubiquitous. Workflow processes are defined for purchasing equipment and supplies, for approving project plans, routing travel expense claims — in short, just about anything that requires two or more persons, groups, departments, or businesses to cooperate.

With the advent of the Web, the ability to perform such workflow tasks online is compelling. Individuals, teams, and business units now communicate electronically. In an e-business[1] world, you can share not only static information via traditional Web pages — you can also experience a global dynamism that encompasses information flow that you can act on.

No doubt, you are reading this book in the hopes of discovering how to effectively implement workflow processes online. Your competitors are doing it. Your suppliers are doing it. Your customers, in vastly increasing numbers, expect it.

In this book, we guide you through the terminology of workflow, and present a framework for writing effective Web-based applications to implement your own business processes. This book is not meant to be an expository introduction to workflow concepts or e-business paradigms. Rather, it is written at a more practical level — we hope to provide process and code examples you can easily extend and modify to develop working applications.

In the next section, we discuss the contents of each chapter. The remaining sections of this chapter explain the standards used in drawing workflow process diagrams, the programming languages and development environments used, and the format of the code examples.

1.1 Chapter Contents

Chapter 2 begins with a discussion of workflow. First, we define workflow and how is it related to traditional business process models. Second, we identify the various component models of workflow. Third, we discuss parallel and serial models of approval processes, and proactive versus passive notification models. Fourth, we illustrate how to plan for and handle exceptions.

In Chapter 3, we introduce Lotus Notes/Domino[2,3] as the software environment for implementing Web-enabled workflow applications. We discuss the specific functions in Notes that enable workflow processes to be implemented. Then, we analyze the standard Notes Approval Template in terms of our understanding of workflow processes. We point out the strengths and weaknesses of this template, in preparation for developing our own framework for Web-based workflow implementations of business processes.

A complete Notes database is developed in Chapter 4. This database design moves us closer to implementation of a generalized workflow template by providing a standalone component to manage notifications. This reusable notification mechanism can be incorporated into existing Notes/Domino applications to manage the automatic generation of e-mail notifications for any combination of users and applications. In later chapters, we reuse this same database as a component of our general workflow framework.

Chapter 5 follows the development of a complete workflow template. We base the workflow elements on a workflow component methodology introduced in Chapter 2, taking advantage of various features of the Notes environment discussed in Chapter 3. We incorporate the notification template developed in Chapter 4 to end up with a reusable framework. We also develop an extensible rules engine to enable the framework to handle automated decisions and set documents to particular states. This framework, complete with all of the code, should enable you to incorporate sophisticated business process modeling in your Notes/Domino Web applications.

We end the book with two chapters that present a case study in workflow implementation. Chapter 6 provides an analysis of a hypothetical college admissions process. We interview the admissions officer, determine the application requirements, and produce a workflow model to describe the application. In Chapter 7, we implement the case study model using our generalized workflow template. The implementation includes a design of the Notes forms and views, customization of the workflow actions, specification of the roles of the people involved, and a review of the resulting process from multiple user points of view.

1.2 Workflow Diagrams

In describing the design of software systems, a standard modeling language has been created. This standard, Unified Modeling Language (UML), was created by Grady Booch, James Rumbaugh, and Ivar Jacobson[4,5]. UML was designed to model various aspects of software systems, from static or structural components, such as classes and interfaces, to behavioral relationships, such as business processes and component interactions.

In this book, we will use two primary diagram types: class diagrams that describe object-oriented classes and relationships among them, and state machine diagrams that model behavior through a series of states and actions that effect changes in state. We will also occasionally use two other modeling diagram types: deployment and activity diagrams.

In this section, we briefly explain these four diagrams without going into all of the intricacies defined in the UML specification. See Fowler[6] for a concise, readable introduction to UML.

1.2.1 Deployment Diagrams

Deployment diagrams model physical systems. Such systems could be computers, distributed processing systems, corporations, server farms, etc. The major encompassing entity in a model of this type is known as a node. Nodes are typically represented by cubes, as shown in Figure 1.1, which defines a "Notes/Domino" server node.

Figure 1.1 A sample deployment diagram showing two component tasks that are implemented in the enclosing node.

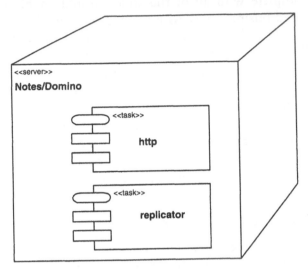

The two shapes labeled "http" and "replicator" represent components. Components represent any of the parts of the node that you wish to model. These can represent individual modules or server tasks (such as the two Notes/Domino tasks depicted in Figure 1.1), specific code modules, databases, user interfaces, etc.

If you examine Figure 1.1 carefully, you will notice a label enclosed in double angle brackets (i.e., "<<server>>") above the labels for the node and the two components. These special labels are designated as stereotypes in UML. Stereotypes allow you to define your own specific modeling language extensions to fit specific modeling needs.

1.2.2 Activity Diagrams

There are two different ways to model workflow processes in UML. The first method we discuss models the activities that take place during a workflow process. UML defines activity diagrams to illustrate what takes place and in what order.

To illustrate an example of an activity diagram, consider a hypothetical workflow process for a scientific journal. In this process, an author writes and submits a paper for publication in the journal. As with most scientific journals, the editor or associate editor selects a set of reviewers that assess the quality of the work, significance to the field, and recommend whether or not to publish the paper. Once all of the reviews are received, the editor evaluates the comments and ratings to decide on either accepting the paper for publication, returning it to the author for revisions, or rejecting it.

We capture this process as a set of activities in Figure 1.2. The black circle at the top of the figure shows the starting point of the process. Each activity is represented by an oval-ended shape labeled with the activity name. Arrows show the direction of the workflow as it progresses from the initial activity of submitting the paper, selecting the reviewers, and having three reviewers return comments to the editor.

Figure 1.2 An activity diagram illustrating the review process for the publication of a peer-reviewed scientific paper.

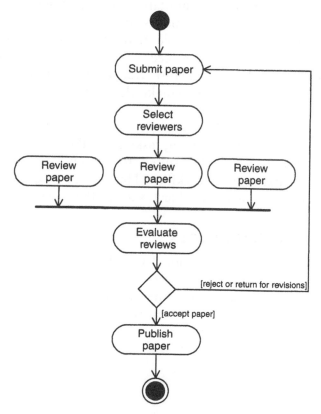

The thick horizontal line below the "Review paper" activities is a synchronization point. It indicates that all three reviews must be received before the process may continue.

Once all of the reviews are received, the "Evaluate reviews" activity results in a decision point, represented by the diamond. The editor can choose to accept the paper for publication, or return it to the author. Note that each path out of the diamond is labeled with a guard condition (denoted by square brackets). The evaluation of the guard conditions as true or false determines the path of the workflow.

Finally, the two concentric circles at the bottom of the diagram indicate the end point of the process.

While this type of diagram does show the activities involved, and the order and conditions under which they are executed, it is an incomplete model. For example, you cannot look at the diagram and tell what individual or group is responsible for handling specific activities or making decisions.

To overcome this limitation, activity diagrams are sometimes redrawn to have swimlanes. Swimlanes are marked at the top of each lane with the name of the actor, or role, that has the responsibility to carry out activities or decisions that are placed in the lane. Figure 1.3 shows a swimlane version of the workflow in Figure 1.2, where we have labeled the roles of author, editor, and reviewer to make clear who participates at each point in the process.

With the ability to specify synchronization points, activity diagrams are quite useful for showing tasks that take place in parallel. However, activity diagrams are better suited to showing the behavior of the actors, or participants, in a workflow. This type of diagram is not best at modeling the state of a particular object, such as the submitted paper in our example.

1.2.3 State Diagrams

The second way to model workflow is to show the state of an object as it moves through the process. This is done by drawing a state transition diagram. We will continue the example from the previous section by showing a state diagram that illustrates the different stages of the process that a submitted paper encounters.

Figure 1.3 A swimlane version of the activity diagram in Figure 1.2, with the roles associated with each activity labeled at the top of each lane.

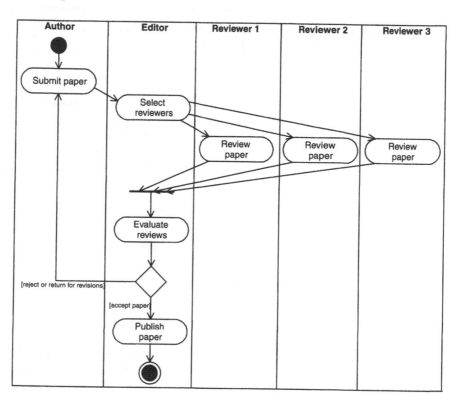

Figure 1.4 A state diagram showing the possible states and transitions allowed by the review process of a scientific journal.

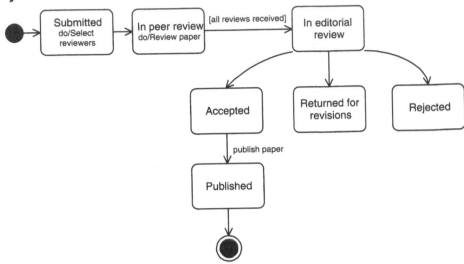

In Figure 1.4, we follow the paper being considered for publication. Each state in the workflow is illustrated by a rectangle with rounded corners. The first state is labeled "Submitted" to indicate that it has been received by the journal.

For each state, we can include an optional set of actions to indicate what must happen while the paper is in that state. In the example, the action "do/Select reviewers" is the principal action that must happen before the paper can transition to the next state. Likewise, when the paper is in the "In Review" state, the "do/Review paper" action is required to be completed.

Note that the synchronization point that is graphically modeled in an activity diagram is specified on a state diagram by a guard condition. In this case, the condition "[all reviews received]" must evaluate to true before the paper can move to the next state.

Once the paper has been reviewed by the editor, it can move to one of three separate states: "Accepted", "Returned for revisions", or "Rejected". Finally, an action label, "publish paper" shows the transition from "Accepted" to "Published". This action may be an explicit part of the process, or may indicate that another process must be initiated (e.g., typesetting, proofing by the author, scheduling, printing, etc.).

Another way to show synchronization of activities before transitions can occur, or to show other subprocesses within a process, is to use superstates.

Figure 1.5 illustrates a superstate labeled "In review" that contains the three concurrent reviewer processes. Once all three of the reviewer processes has completed (i.e., moved to their final state within the superstate), the paper transitions to the "In editorial review" state.

Figure 1.5 A state diagram illustrating substates within a superstate.

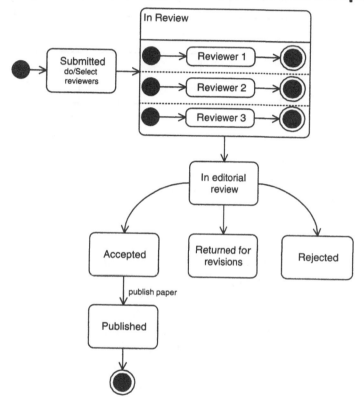

A superstate might also be used to describe a nonconcurrent process, such as the subprocess associated with actually printing and publishing an accepted paper.

1.2.4 Class Diagrams

The fourth common diagram type we use in this book is a class diagram. Class diagrams illustrate the structure of code, particularly for object-based or object-oriented designs. For example, consider how to model the state transitions of an object like the manuscript of our workflow example. We might choose to use a class structure like the one shown in Figure 1.6.

Figure 1.6 A typical class diagram illustrating the State design pattern[7,8] to track state changes of an object.

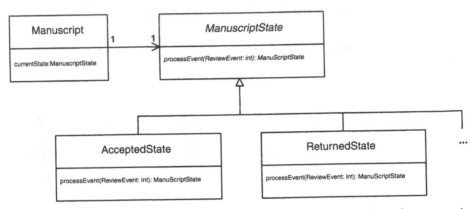

Classes are shown as rectangles, with different kinds of connections between them. For example, the *Manuscript* class models instances of papers. Each paper has an instance variable called *currentState* that is of type *ManuscriptState*. The arrow between these classes indicates an association between them. The direction of the arrow indicates that papers know their state, but that states know nothing about their associated papers. The numbers above the arrows denote that there is one state for each paper (i.e., a 1:1 relationship).

The *ManuscriptState* class and its associated method *processEvent()* are shown in italics because they represent an abstract class and method, respectively. Thus, instances of *Manuscript* can deal with the abstract data type and method, but the actual or concrete type and method are implemented in derived classes of *ManuscriptState*. Listings 1.1 and 1.2 show how these classes would be represented in Java.

Listing 1.1 The *Manuscript* class.

```java
public class Manuscript
{
    // other methods and properties
    // ...
    private ManuscriptState currentState;
}
```

Listing 1.2 The *ManuscriptState* **abstract class.**

```java
import java.awt.event.*;


abstract public class ManuscriptState
{
    // other methods and properties
    // ...
    abstract public ManuscriptState processEvent(ActionEvent e);
}
```

The derived classes are shown in Figure 1.6 as *AcceptedState*, *Returned-State*, etc. These are connected to the base class with an open arrowhead connector to indicate that they are derived from a generalized, or base, class.

The individual *processEvent()* methods in each derived state contain the business logic that determines, based on events received, into what new state the manuscript may transition. Listing 1.3 shows how the *Accepted-State* class would be implemented in Java.

Listing 1.3 The *AcceptedState* **derived class.**

```java
import java.awt.event.*;


public class AcceptedState extends ManuscriptState
{
    // other methods and properties
    // ...
    public ManuscriptState processEvent(ActionEvent e)
    {
        // business logic for processing ReviewEvents
        if ( e.getActionCommand().equals("PUBLISH") )
            return new PublishedState();
        else if ( e.getActionCommand().equals("REJECT") )
            return new RejectedState();
        else
            return ReturnForRevisionsState();
        return new AcceptedState();
    }
}
```

Figure 1.7 shows another part of a class structure model for our work-flow example. In this case, an observer pattern[7,8] implementation is shown. An *Editor* class implements an interface named *ReviewEventListener*. The implementation of an interface is diagrammed as a dotted line with an open arrow pointing to the interface.

Figure 1.7 An implementation of an Observer design pattern for receiving notifications of events.

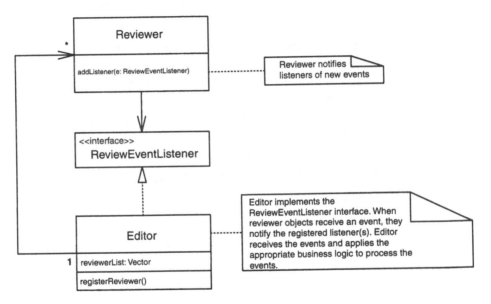

Two other things to note about Figure 1.7 are:

- the annotation blocks that describe the connections among the three classes; and

- the association between Editor and Reviewer, which indicates a one to many relationship (i.e., 1:* — an editor can use any number of reviewers).

Listings 1.4 and 1.5 show the definition of an interface and a class that implements that interface.

Listing 1.4 The *ReviewEventListener* interface.

```
public interface ReviewEventListener
{
    public abstract void reviewReceived(ReviewEvent e);
}
```

Listing 1.5 The *Editor* class implementing the *ReviewEventListener* interface.

```
import java.util.*;

public class Editor implements ReviewEventListener
{
    // registerReviewer
    public void registerReviewer(Reviewer r)
    {
        // add this instance to the list of listeners
        // for review events
        r.addListener(this);

        // add reviewer to private list
        reviewerList.addElement(r);
    }

    // implementation of ReviewEventListener
    public void reviewReceived(ReviewEvent e)
    {
        // handle events
        // ...
    }

    // properties
    Vector reviewerList = new Vector();
}
```

1.3 Environments, Languages, and Code

The workflow management tools, templates, and code discussed in this book are based on the Lotus Notes/Domino environment. All of the Notes/Domino examples were created on the version of Notes known as R5, using the Domino Designer application. The Web-based example programs were tested on either Windows 95 or Windows 98 client machines, running either Netscape Navigator version 4.6+ or Internet Explorer version 5.0+. The client machines were tested against server applications running on Windows NT Server version 4.0 with service pack 5, running Domino version 5.0.1 or greater.

The Notes agents that implement much of the workflow management system's functionality were developed in either of two languages. The first language, LotusScript, is the BASIC-like built-in scripting language for Notes. Note that LotusScript does allow the creation of classes in the object-oriented sense. In the latest release of Notes/Domino, R5, the language added an inheritance mechanism. Before that, LotusScript was at best an object-based rather than an object-oriented language.

The second agent language is Java, which provides a fully object-oriented language with support for many third-party packages. It also provides full access to features such as abstract base classes and interfaces. As such, the Java version of the agents may differ substantially in design from a LotusScript implementation. However, both languages utilize the same underlying interface to Notes databases, using the Notes Object Interface (NOI) classes.

All of the code in this book is included on the accompanying CD-ROM. The code examples are organized by chapter, with each chapter having its own corresponding subdirectory. Table 1.1 shows the naming conventions followed for the files on the CD-ROM.

Table 1.1 CD-ROM file naming conventions.

File Type	Naming Convention
LotusScript agent source file	*name*.ls (e.g., test.ls)
Java agent source file	*name*.java (e.g., test.java)
Java agent class file	*name*.class (e.g., test.class)
JAR file (Java archive)	*name*.jar (e.g., test.jar)
Notes template	*name*.ntf (e.g., notify.ntf)
Notes database	*name*.nsf (e.g., notify.nsf)

In the text, references to variables or method names are shown in italic monospaced font (e.g., `wfAgent`). All code examples are shown in a monospaced font (e.g., `x += 1;`).

1.4 Summary

In this chapter, we summarized the contents of the chapters to follow. UML diagrams were introduced as a standard modeling language to describe workflow processes. Also, we explained the development environment and programming languages used to create the workflow management system.

In the next chapter, we introduce the general concepts behind workflow management systems. This will provide a basis for the implementation details presented in the rest of the book.

1.5 References

1   IBM e-business. [http://www.ibm.com/ebusiness]

2   Lotus Development Corporation. [http://www.lotus.com]

3   Lotus Development Corporation. Notes.net. [http://notes.net]

4   James Rumbaugh, et.al. 1999. *The Unified Modeling Language Reference Manual*. Reading, MA: Addison-Wesley.

5   Grady Booch, et.al. 1999. *The Unified Modeling Language User Guide*. Reading, MA: Addison-Wesley.

6   Martin Fowler with Kendall Scott. 1997. *UML Distilled: Applying the Standard Object Modeling Language*. Reading, MA: Addison-Wesley.

7   Erich Gamma, et.al. 1994. *Design Patterns: Elements of Reusable Object-Oriented Software*. Reading, MA: Addison-Wesley.

8   Mark Grand. 1998. *Patterns in Java*. Volume 1. New York: John Wiley & Sons.

Chapter 2

Workflow

One of the revolutionary innovations of the twentieth century manufacturing industry was Henry Ford's vision and implementation of the assembly line. Ford devised this solution as a way of keeping up with increased demand for the Model T automobile in the early 1900s. Previously, the production steps necessitated each worker coming to the car to perform their tasks as it was being assembled. Some workers performed many different tasks, which prevented them from attaining higher levels of proficiency and efficiency. Mr. Ford theorized that having the worker remain in place, and assigning them one specific task to do as the automobile came to them would speed up manufacturing time and decrease the total number of man-hours. Thus, both production and profits would increase.

Ford tested his idea by dragging a chassis by a rope and windlass along the floor of the Highland Park, Michigan plant in mid 1913. The result was a revolution in modern mass production. Subsequently, Model Ts popped from the assembly line at the rate of one every ten seconds.

True to their recent advertising campaign, Ford did have a better idea!

As much as Ford's idea revolutionized the manufacturing industry for the rest of the twentieth century, he also unknowingly designed one of the first automated workflow systems.

However, Ford did not discover the basics of workflow. Systems have existed since man first stumbled on the need for productivity through a defined process. Manual and paper-based systems were designed and refined by pocket-protector-armed Efficiency Experts, the forerunner of today's system analysts. The computer first revolutionized workflow by automating some of the "work" items in various processes, and subsequently addressed the "flow" in workflow with the introduction of in-house networks and electronic mail.

We are again in revolutionary times. The new millennium sees our industry and the world in the midst of what seems like a combination of the Industrial Revolution and the California gold rush combined! Indeed, businesses are reinventing, redesigning and reengineering themselves, with millionaires being made overnight, all because of the Internet.

> Every day it becomes more clear that the Net is taking its place alongside the other great transformational technologies that first challenged, and then fundamentally changed, the way things are done in the world.
> Lou Gerstner — Chairman, IBM Corporation

We are entering an age of "market facing systems," which is turning the terminal or monitor away from the employee in the cubicle, to face and be used by the outside world. The assembly line is expanding out of a building or company, and into a space that is limited only by the existence of a connection to an Internet Service Provider. This worldwide infrastructure, with "next to nothing" deployment costs, creates a market and platform that has never existed before. Although estimates change daily, it is forecast that the Internet will connect over 600 million users by the year 2001. With technologies such as Lotus Notes/Domino, we are not limited by proprietary systems and software or by closed networks in connecting with that user base.

Workflow[1] is quickly becoming the assembly line technology of the Internet. Instead of a skeletal car moving from worker to worker for the next part to be installed, documents are appearing in e-mail in-boxes for approval. Items purchased online are shipped and tracked from origin to destination. Application forms for just about anything are processed, all with the next appropriate person or system receiving the work on a virtual assembly line.

2.1 Workflow — What is it?

Any study of workflow for the millennium cannot be started without a review of some basic components. These components will not change, regardless of the supporting technology.

We urge you to not skip over the following foundation explanations and go right for the code examples. Remember, the process and the workflow that supports it are more important than the enabling technologies. As such, if you don't fully understand and possibly reengineer a process, the resulting workflow and the software that enables it will be disabling and counter-productive. Often, projects linger for years because a developer jumped right into coding to a design in his head, leaving process issues secondarily addressed. While this may be an advantageous situation for a consultant charging by the hour, it is generally inadvisable to have development projects last longer than presidential terms.

Our first building block is the first step in any workflow analysis: process. As this term is too broad and all encompassing, it helps to refine it for our discussion as *Business Process*.

The Workflow Management Coalition (WfMC) defines a business process as:

> A set of one or more linked procedures or activities which collectively realize a business objective or policy goal, normally within the context of an organization structure defining functional roles and relationships.
>
> *Workflow and Internet: Catalysts for Radical Change*[2]

Andrzej Cichocki, et. al., define a process as:

> A description and ordering of work activities across time and space that is designed to yield specific products and services while ensuring the organization's overall objectives.
>
> *Workflow and Process Automation — Concepts and Technology*[3]

2.1.1 Process

Q. What is a process?
A. Everything is a process.

If we analyze the WfMC definition of *process*, we quickly see that every goal or objective of a company is achieved by a series of one or more, possibly interrelated, processes.

To view a company from a process perspective, let's perform some process decomposition on the fictitious Acme Widget Company (Figure 2.1). Acme's main business goal is to increase stockholder value through designing, manufacturing, and selling world-class widgets.

Figure 2.1 The organizational structure of an example company.

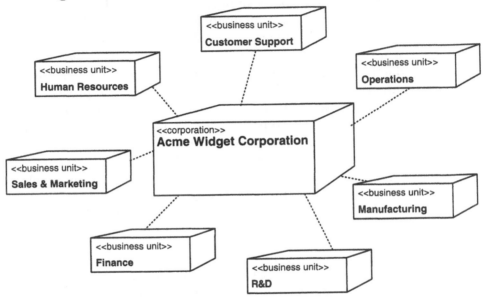

If we start at the main business goals of designing, building, and selling widgets, we can follow the logical decomposition of the main processes. These processes are fairly generic, so they should be common to most businesses.

For example, we may analyze the processes for one of the business units in Figure 2.1, the Manufacturing unit. Figure 2.2 shows a number of processes that the manufacturing unit must implement to carry out its mission.

Each of these single processes must be further defined by identifying the subprocesses and dependencies associated with that process, roles (human and system), inputs needed, and outputs expected.

Analysis, understanding, and possible reengineering of the business process are necessary before a workflow system can be designed. Any missing elements or inefficiencies in the process modeling or definition will become black holes in an automated workflow system. At a minimum, this will create increased exception handling considerations or, even worse, a workflow design that cannot be implemented.

Figure 2.2 The processes associated with a business unit.

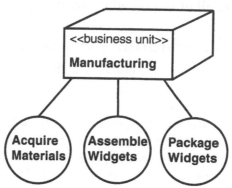

The abstraction or model of the business process serves as the starting point for a workflow design.

2.1.2 Why Should I Model?

To paraphrase M. Scott Peck[4], "Software is difficult." The complexity of modern software and systems seems to increase in proportion to the advances and functionality breakthroughs of computer science and technology. Although you might react to the previous sentence with a resounding "Duh!", considerable effort has been expended in the last decade to simplify systems development. Now we build tools and systems that do not require multiple advanced degrees as prerequisites for usage, but rather enable our continued meteoric advancement in software and systems design.

The reality is that people have always needed the benefits of modeling to enable our understanding of systems, as our ability to comprehend complexity is limited. As the UML triumvirate (Booch, Jacobson, and Rumbaugh) stated:

> We build models of complex systems because we cannot comprehend such a system in its entirety.
>
> *The Unified Modeling Language User Guide*

They also identify four aims that we achieve through modeling:

- Models help us to visualize a system as it is or as we want it to be.
- Models permit us to specify the structure or behavior of a system.
- Models give us a template that guides us in constructing a system.
- Models document the decisions we have made.

These are certainly all-important goals in constructing a workflow system with all of its potential complexity.

2.1.3 How Do I Model a Process?

So now that you are convinced that we need to model our processes, your next question is probably "How?" As with most things, the answer is usually, "it depends."

If you are a user of, or dictated to use, a particular systems development methodology, you may have a specific set of modeling tools and types that you use. In this section, we review a few basic modeling methods to acquaint those new to this activity. Those readers already experienced in modeling may want to skip to the next section, "UML Models" on page 26.

This is by no means a primer on process modeling or a review of all available tools and methodologies. Modeling is simply an important building block in our construction of a Domino-based workflow system that will help us visualize, specify, guide, and document the system.

If you venture into the world of process reengineering and workflow, just about any book has a version of a diagram for communications-based workflow (Figure 2.3). The communications-based methodologies are derived from the work of Winograd and Flores, who developed the Conversation for Action Model[5].

Figure 2.3 A communications-based model of workflow.

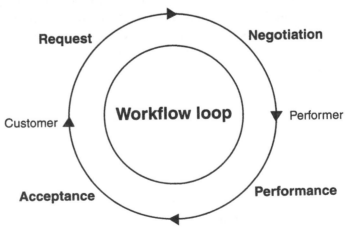

This communications model is based on at least two sets of actors, the Customer and the Performer, and four main phases.

Request   The customer initiates a request.

Negotiation   The customer and performer negotiate the request and come to agreement.

Performance   The performer executes the requested action.

Acceptance   The customer communicates acceptance when an action is completed to satisfaction.

To start with a sort of "Hello, world" example of modeling, we pick a non-computer-oriented process familiar to most people. Our example — buying a hot dog from a street vendor — is not only familiar, it's also one of life's simpler pleasures, and probably one of life's riskier activities. High anxiety and high fat — you can't beat it. Let's model that process using the same type of communications-based model shown in Figure 2.3. The result is shown in Figure 2.4.

Figure 2.4 A communications-based model for buying a hot dog.

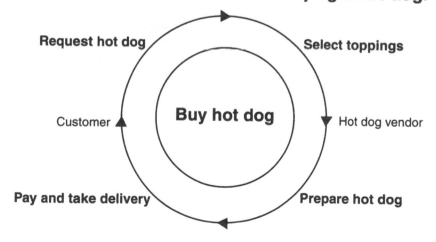

Our request phase is the customer's verbal request to the vendor for a hot dog. The negotiation phase might include a "Whadda ya want on it?" query from the vendor, or a "How much is that?" query from the customer. Our process enters the performance phase as the vendor prepares our culinary delight, and we finish with the acceptance phase as the customer forks over the money and takes delivery of the delectable morsel.

With the exception of the performance phase, each state involved two actors, one in the customer role and one in the vendor role. Each of our process loops can, and probably does, involve secondary loops. These secondary loops may be required for the main loop to reach completion. They will continue to expand until they are irrelevant to the goals of our particular process. Let's pick the "Prepare hot dog" performance phase from Figure 2.4 and decompose it.

Before the vendor can prepare the hot dog, he must (hopefully) cook it. That process itself can be broken down into many activities. In many of these activities, the vendor is still the performer, but other participants may act as additional performers. That process is more appropriately modeled with an activity-based model, such as the one in Figure 2.5.

Figure 2.5 Activity diagram that models the performance phase of Figure 2.4.

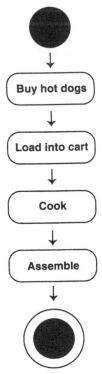

Before the vendor cooks the hot dogs, he must buy them. That process sees a role reversal, where the vendor becomes a customer who enters into a workflow loop that resembles our original Conversation for Action Model, but with differing roles. This is shown in Figure 2.6, in which the hot dog supplier has taken the "vendor" role. Note that decomposing a process exposes subprocesses that may need to be modeled.

Figure 2.6 Role reversal for the vendor, who becomes the customer in a subprocess.

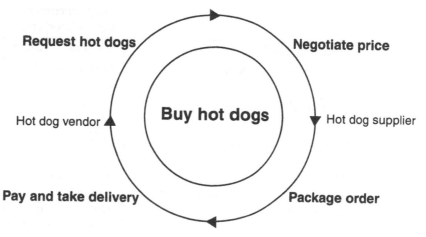

Now that we have some very basic processes modeled, what do we do with them? Well, models are meant to serve as the basis for testing and analysis. We must scrutinize the model and answer the following questions.

- What are the relevant subprocesses?
- What are the dependencies?
- What triggers the process?
- Who are the participants?
- What are the inputs?
- What are the outputs?
- What are the business rules, implicit or explicit, that define, support, or limit the process?
- What determines whether the process is a success?
- What exceptions or errors may occur?

We complete our hot dog purchase process by answering the previous questions in Table 2.1.

Table 2.1 Standard questions to scrutinize a process model.

What are the relevant subprocesses?	Buying hot dogs (wholesale) Cooking hot dogs
What are the dependencies?	Cooked hot dog Condiments Supplementary materials (cart, napkins, plates, etc.)
What triggers the process?	Customer request
Who are the participants?	Hot dog vendor Customer
What are the inputs?	Detailed request by customer Money received from customer
What are the outputs?	Hot dog prepared to customer specifications
What are the business rules, implicit or explicit, that define, support, or limit the process?	Materials (hot dogs, buns, etc.) must be fresh Hot dogs must be cooked for five minutes in 350-degree clean water Cooked hot dogs must have a temperature of 120 degrees
What determines whether process is a success?	Satisfied customer
What exceptions or errors may occur?	Customer changes or cancels request Hot dogs not fully cooked or improperly prepared

As you can see, even the most ordinary and mundane of daily activities can be an involved, multifaceted process when examined closely.

2.2 UML Models

To design a workflow for our example process, we must visually diagram the process and all of its characteristics as identified previously. No single diagram can be drawn to properly illustrate the analysis with all of its rules, states, decision points, etc. Thus, we use a couple of standard diagrams to illustrate different perspectives on the process.

First, look at a UML activity diagram of the hot dog purchase process (Figure 2.7). From this diagram, we see our major phases (request, negotiation, performance, and acceptance) translated into eight states with three decision points. However, as noted in Chapter 1, a drawback of this diagram is that it is unclear who is performing each activity.

Figure 2.7 An activity diagram for the hot dog purchase process.

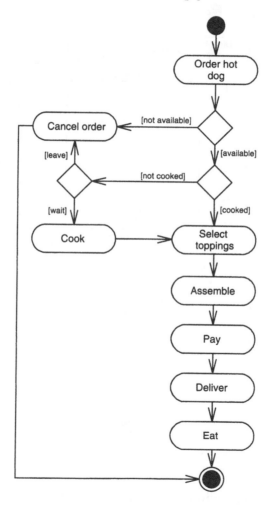

As discussed in Chapter 1, swimlanes provide a way to graphically represent the major roles, persons, or entities responsible for a particular action. A swimlane activity diagram provides an excellent introduction to a process by providing both the flow of activity and the involved parties. Figure 2.8 shows the swimlanes added to the process diagrammed in Figure 2.7.

Figure 2.8 A modified activity diagram that includes swimlanes to indicate the roles that take part in the process.

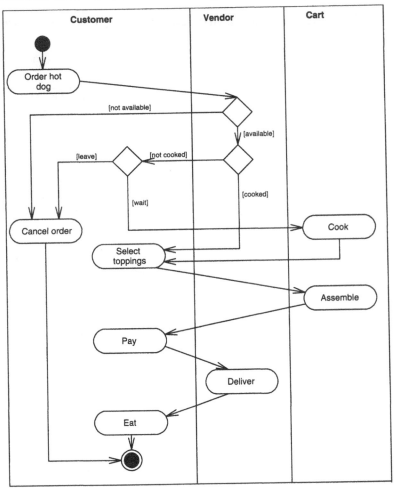

In our swimlane activity diagram, we introduce a new role — the hot dog cart. We include the cart because it is part of the overall process, and it illustrates how roles in a workflow process are not always persons or departments. Roles can be spreadsheets, databases, accounts payable systems, stand-alone computers, or even hot dog carts.

Another viewpoint critical to our workflow design capabilities, especially in the document-based Notes/Domino environment, is a state diagram. Here our hot dog example somewhat fails us, unless we consider the hot dog itself as a "document." Then we can assign to it a series of states that describe the hot dog throughout the purchase process.

Figure 2.9 shows a state machine diagram for a purchase request, illustrating the various states that the request assumes during its lifetime. We see the request being submitted, the condiments selected, the assembly process underway, completion of the order, and payment of the order. Action labels that appear near the line between some of the states (e.g., pay for order) are known as event triggers. An event trigger moves the request from the source state to the target state, once all conditions or business rules have been met.

Figure 2.9 A state diagram for the hot dog purchase process.

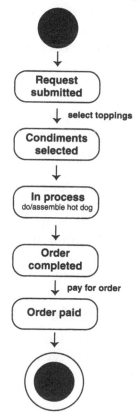

2.3 Evaluating the Model

Now that our process is modeled from several perspectives, we should review the models with the following four situations in mind.

Reach   Can the workflow process reach the required state?

Safe   Will an exception to the process leave the workflow in an unacceptable state?

Deadlock   Can a state be reached that allows no activities to be performed?

Conflict   Do two or more activities require a common resource?

While these qualitative characteristics are being analyzed, the quantitative characteristics must also be identified (e.g., time frame necessary to achieve a goal, potential bottlenecks, etc.).

2.4 Methodologies

Most workflow methodologies have the main components or building blocks of workflow packaged in such a way as to make them easy to remember. Some familiar ones are Roles/Rules/Routes and Actors/Actions/States.

These are good frameworks for approaching workflow, but can be somewhat limiting. We present an expanded list of workflow components and considerations with a brief description of each one. Consider these methodology descriptions when designing or redesigning a workflow system, and use them as your building blocks. No doubt, you will be able to extend and refine them further, as you should, for your own organization or customer.

2.4.1 Actors, Roles, or Performers

Any person, group, organization, computer system, database, file, program, agent, or macro can initiate, trigger, request, supply, contribute, finalize, or effect a workflow process.

This would constitute the Who, and occasionally the What, in a Who-What-Where-When-Why form of analysis. Actors are best abstracted into classes to allow for system flexibility and growth. A shortsighted approach is to identify specific individuals involved and then allow for replacement of those persons. This builds in rigidity to the role that will not allow for future larger scale process changes.

A top-down object-oriented approach allows for greater flexibility and maintenance. This approach requires you to identify a general set of classes for the potential actors in a workflow process. Then you refine their specific roles through the properties and methods of those classes.

2.4.2 Routes, Paths, or Flow

These words describe the journey taken by a process or document as it fulfills its intended reason for being.

The amount of structure in routing will depend on the type of workflow process with which you are working. An ad hoc system, such as proposal production or a collaborative design process, will offer very flexible and user-controlled routing. At the opposite end of the spectrum, a benefits enrollment or help-desk process may be more structured because of the predictability of the process and limited options available to the participants.

2.4.3 Actions, Events, or Triggers

These nouns refer to the dynamics of a process. All workflow processes are dynamic by nature, and actions provide that movement or flow. An action can be something as obvious as an employee submitting an expense voucher, or as subtle as the expiration of a time period for approval of that same voucher.

2.4.4 Exceptions and Guard Conditions

In UML terminology, a guard condition is a piece of logic that must be satisfied before an action can occur. Guard conditions can be thought of as a modeling implementation of business rules. The importance in identifying them in the model is to provide for alternative actions if the conditions are not satisfied. If this is not done, a deadlock condition may occur.

2.5 Business Rules

As processes define an organization's method of achieving its business goals and purpose, business rules define the policies and constraints that form those processes. Rules provide the knowledge upon which the processes rely.

Although rules are sometimes inappropriately formulated in the software design and redesign process, they are truly the possession and responsibility of the business process owner who must define and implement them.

An analysis of a system not performing up to expectations frequently reveals that the business process owners were not included in all phases of the design and development of the system. The result is usually an inflexible system caused by hard-coded or scattered business rules. Rules will change, and a properly designed workflow process should provide for painless modification.

2.6 Communications

While any communication that takes place during a workflow process could accurately be classified as an action, we are going to separate this category because of the array of types and methods of communications that potentially play a part in a Web-based workflow process.

In addition to the type of communication (e-mail, fax, phone, workgroup in-basket item, etc.), we need to consider the method of communication.

For instance, a notification method can be proactive or passive. A proactive method would be the generation of an e-mail, fax, or phone message to the next actor in a workflow process. The method may allow repeat notifications if no action is taken within a specified period of time. A passive method could be simply the deposit of a work item in a queue for the worker to discover when they next checked their to-do lists. A method could even be so passive that there is no notification whatsoever, but merely a dependence on the state change itself to be the communication.

Actions that must be performed by numerous people when a document is in a consistent state, such as multiple approvals for a submitted application, can be accomplished in a parallel or serial fashion. A parallel scenario would have the document presented to all intended approvers at the same time, letting them simultaneously process the document. A serial approval design would present the document to each approver in turn, with the next approver receiving the document only after the previous approver has processed it.

2.7 Summary

Workflow as a technology has the potential to be as important to the Internet age as the assembly line was to the manufacturing industry. The strength of an efficient workflow design is grounded in a clear understanding and possible reengineering of the underlying processes. Workflow-based implementations of processes are complex — developers must model the processes to attain a clear understanding of all aspects of a design.

The building blocks of workflow (States, Routes, Actions, Actors, etc.) are characteristics of the workflow process that are identified through proper analysis and design of the workflow.

The preceding process and workflow study is independent of the technologies that will be used for their implementation. In Chapter 3, we review the rich set of functionality that Notes/Domino offers, and introduce how to take advantage of this functionality in workflow applications.

2.8 References

1 F. Leymann and D. Roller. 2000. *Production Workflow: Concepts and Techniques.* Upper Saddle River, NJ: Prentice-Hall PTR.

2 Workflow Management Coalition. *Workflow and Internet: Catalysts for Radical Change.* [http://www.aiim.org/wfmc/mainframe.htm]

3 Andrzej Cichocki, et.al. 1998. *Workflow and Process Automation: Concepts and Technology.* Norwell, MA: Kluwer Academic Publishers.

4 M. Scott Peck, M.D. 1978. *The Road Less Traveled.* New York, NY: Simon and Schuster.

5 Terry Winograd and Fernando Flores. 1997. *Understanding Computers and Cognition.* Reading, MA: Addison-Wesley.

Chapter 3

Workflow Using Notes/Domino

Notes/Domino is frequently included in articles and white papers on workflow. However, it is quickly described and dismissed as "...groupware, not a workflow management system," which is correct.

Some of the more complimentary articles have described Notes as a platform for workflow, which is also correct. Out of the box, Notes/Domino does not offer true workflow functionality.

Several Notes/Domino templates, which we'll review in more detail later in this chapter, do provide us with a specific type of limited workflow capability. However, they are only starting points for the kind of Workflow Management System (WFMS) that is possible with Notes/Domino applications.

Until Lotus' purchase of ONEStone's Prozessware, Notes/Domino was missing the design and execution components necessary to make Domino a WFMS. Prozessware will bring to Domino a workflow architect, engine, and viewer. This acquisition is a natural "missing piece of the puzzle" addition for Lotus to provide a full-blown WFMS to large corporate clients.

The principal mission of this book is to present the rich workflow potential of Domino, as is, and craft a simplified and extensible workflow management system from scratch. This book serves as a guide, blueprint, and example for coding workflow applications in the small and medium-sized business arena, where the economics of high-end solutions are not warranted.

3.1 Notes Features

Notes/Domino implements numerous design elements and features that can be used independently or in combination to create powerful, secure, workflow-enabled Internet or intranet applications. In the following sections, we review some of the traditional features of the Notes/Domino product and highlight their workflow potential. We conclude this chapter with a review of the R5 version of the standard Document Library template. This template illustrates some of the basic workflow functionality of Notes.

Figure 3.1 Notes workflow-enabling features.

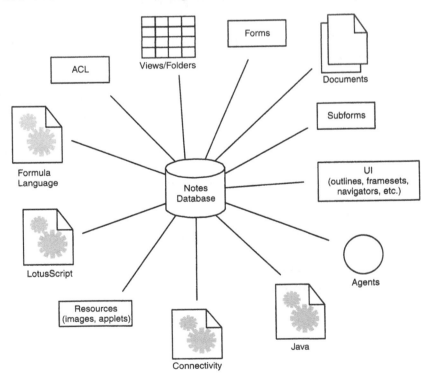

3.1.1 Notes Security

Security for a workflow application and its users is a critical element for accomplishing the objectives of a business process. Our concerns in this regard are two-fold: accurate identification of the users and controlled access to the application functionality and data. Fortunately, the robust security model of the core Notes client/server design is inherited for our Internet application design tasks.

3.1.1.1 User Security

Browsers cannot use the Notes client user ID file to authenticate users. In its place, Domino supports varying levels of user security. First, basic HTTP username and password authentication is supported. Second, Domino R5 provides an option for form-based login and persistent server sessions (using cookies) that can automatically time out.

Domino also supports multiple levels of Secure Socket Layer (including SSL3 Client Authentication) and certificate authentication (e.g., X.509 digital certificates).

3.1.1.2 Architectural Security

The Domino security model offers six layers of control. This model gives us a logical, progressive, outside-in approach to granting access to data or application functionality. Through HTTP basic authentication, we can grant or deny user access ranging from the server level to the field level of individual documents.

This level of control offers powerful capabilities to our workflow design options. Table 3.1 presents the layers with specifics on how each layer may be secured.

Table 3.1 Security options and access layers of Domino security layers.

Security Options	Access Level
Server	
Binary permission set in the server document	
Allow anonymous connections in the security section of the server document.	Grant access to anyone.

Table 3.1 Security options and access layers of Domino security layers. (continued)

Security Options	Access Level
Allow server access only to users listed in this Domino directory (Yes/No).	Grant access only to registered users.
Define who can access the server.	Grant access to specific people, groups, and servers.
Define who cannot access the server.	Deny access to specific people, groups, and servers.
Define who can create new databases.	Grant access for specific people, groups, and servers to create new databases on the server.
Define who can create replica databases.	Grant access for specific people, groups, and servers to create replica databases on the server.
Database	
Define access control list (ACL).	Grant access to specific users, groups or servers, and control read/write/delete functions for user or group.
Form	
Define default read access for documents created with this form.	Grant access to all readers and those with higher access, or to specific groups and individuals.
Define who can create documents with this form.	Grant access to all authors and those with higher access, or to specific groups and individuals.
Define the Paragraph Hide When formulas to control the display of specific lines or paragraphs in a form.	Grant access to specific paragraphs based on the mode of a form (read or edit), or on the result of a formula. (e.g. @If Status = 'New').
Define readers names fields.	Grant access to specific users or groups who can read a document, but do not override the database ACL level.

Table 3.1 Security options and access layers of Domino security layers. (continued)

Security Options	Access Level
Define authors names fields.	Grant access to specific users or groups who can create a document, but do not override the database ACL level.
Document	
Once a document is created via a form, restrict or allow access to the specific document with the Readers and Authors names fields, if they are editable.	Grant access based on Readers and Authors fields.
Section	
Control section access by a formula that identifies users who can edit fields in the section.	Grant access based on a formula.
Field	
Restrict access to users with at least Editor access to enter data.	Grant access based on a formula.
Control the display of specific fields in a form with Paragraph Hide When formulas.	Grant access based on a formula.

3.1.2 Notes Databases

Q: When is a database not a database?

A: When it is a Notes database.

If we could turn back the clock to 1984 when Ray Ozzie, et. al., started creating what we now know as Notes, we might offer one suggestion. We would suggest that the central Notes container object should not be called a database. This term causes confusion for those new to Notes technology because of preconceived notions of what constitutes a relational or object-oriented database.

Databases, in their traditional technical context, are computing models for data. They organize fields of data into rows, columns, tables, or objects. Complex relationships can be formed with other tables or databases to solve queries posed via an application interface, report writer, or simple

Structured Query Language (SQL) statement. Also, databases traditionally allow for the separation of data from the applications that utilize it.

By contrast, Notes/Domino is neither an object-oriented database management system (OODMBS) nor a relational database management system (RDBMS). Notes started as and remains a document-based technology. Instead of a database consisting of many tables with data in structured rows and columns, a Notes database contains unstructured data in documents. Instead of extracting data from a traditional database with an SQL statement, Notes data can be selected and viewed via forms for individual documents, or via views for collections of documents. This document-based design lends itself perfectly to workflow technology, which is predominantly document-based.

In a Notes/Domino application, the data and application code may coexist in the same database. Indeed, for many Notes technologists, the terms "database" and "application" are synonymous. One physical Notes Storage Format (NSF) file can contain all of the data and program code to enable a fully functioning application.

When teamed with the messaging capabilities of Domino, and the other functionality of the Notes/Domino architecture, we have a set of technologies and tools that lets us build, and continue to evolve, powerful workflow applications for the Internet.

Let's look at the elements that make up a Notes database. This review is not meant to serve as a technical description or reference on Notes design elements. Rather, it is a brief review of the rich set of Notes capabilities from a workflow perspective.

The Notes database itself offers us much in the way of workflow functionality. In addition to the database security functionality already mentioned, the application-specific tool of roles is available. Roles may be thought of as a subset or refinement of a database's ACL. If a group does not exist in the Domino Directory for your intended usage, and this collection of people only needs to be identified for a single application, you can create a role and assign individuals or other groups to the role. Roles can be used in formulas to provide flexibility in structuring access control without requiring access to the Domino Directory.

3.1.3 The Notes Object Store

The Notes object store is unique. It is a self-contained, searchable, dynamically sized repository for all of the data, attachments, and application code in a Notes application. It provides a flexible container for all types of data (text, video, audio, binary and text file attachments), and even supplies a functional platform[1] for the creation and streaming of media content (e.g., streaming of RealVideo files via HotMedia Connect for Domino).

3.1.4 Views

Views are a window into a Domino application's data. Views are used both as part of an application's user interface and to programmatically select subsets of documents. As a workflow tool, views provide a customizable, secure interface for the organization and presentation of selected documents. Views can be used for basic monitoring of workflow processes, or used as work queues for the management and disbursement of tasks to a group of users. As a programmatic tool, views give us the ability to select subsets of documents to process.

3.1.5 Documents

A Notes document is an entry in a Notes database, a record in the parlance of the traditional database world. Documents consist of one or more fields that contain many types of data.

A document can be of varying size and can grow or recede in size programmatically during the course of its existence. A document is not strictly comparable to a hard-copy document in that you can't physically view a document by itself. For that, you need a form. You can access, update, or delete the data in a document programmatically, using event triggered scripts or agents, or from the event structure available in forms.

3.1.6 Forms

Forms are templates used to view all or a portion of a document. They can give context and perspective to the underlying data. Forms offer flexibility by enabling multiple customized interfaces and functionality for the data within a document.

A workflow-enabled process may consist of a single document but utilize multiple forms to view that document throughout its lifecycle. For example, consider an online mortgage application such as the one depicted in Figure 3.2.

Figure 3.2 An online mortgage application and servicing process.

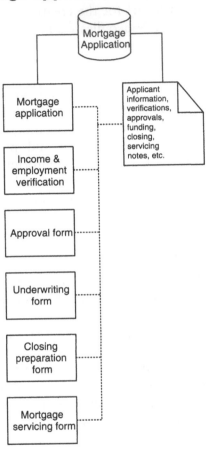

Once the basic applicant information is collected, the document is forwarded to a department within the mortgage company for validation of employment and income. When those verifications have been satisfied and the loan is approved, the document is sent to an underwriting group to fund the loan. After the loan is funded, the materials are prepared for the actual closing of the loan at settlement. When the loan closes, the document is transferred to the servicing department for monitoring and maintenance of the loan.

While all of the same data is not needed at every step in this process, much of the document content is needed by various departments, albeit framed differently for each department's specific purpose. Multiple forms could be designed that display or manipulate just the data pertinent to a specific step in the process.

3.1.7 Subforms

A subform is a smaller version of a form that can be embedded within a form. Subforms are advantageous from a design perspective as we can maintain the same collection of fields for re-use in numerous other forms without re-creating the fields, logic, etc. Also, subforms can be inserted programmatically into a form to customize portions of the form for specific users, groups, or conditions.

A simple example of this feature would be an intranet application that lets the employees of a company perform their annual selection of benefits (medical, dental, etc.). When a user selects a specific level of medical coverage, a subform is programmatically inserted. The subform contains a list of the medical providers offering that level of coverage. System performance is maximized because the size of the form is greatly reduced. Usability is improved because the employee only views providers pertinent to their coverage level selections.

3.1.8 Fields

There are fourteen field types in Domino R5. Each type of field has numerous properties and methods that can be set through the Notes client, Designer, or programmatically through formulas or agents. Table 3.2 lists the most common field types along with their workflow capabilities.

Table 3.2 Notes/Domino field types.

Field Type	Workflow Considerations
Text	Text fields, like most field types, can utilize Hide When capabilities
Date/Time	Most workflow processes are time managed or time dependent. This field type permits the display of a date/time in a variety of formats. Dates may range from 1/1/0001 through 12/31/9999 (for those long workflow cycles). Entering two-digit years between 00 and 49 assumes the century starting in the year 2000. Entering two-digit years between 50 and 99 assumes the century starting in the year 1900. Times range from 00:00:00:00 through 23:59:59:59 in 24-hour format and from 12:00:00 AM through 11:59:59 PM in 12-hour format.
Number	Standard numeric field functionality along with Hide When capabilities.

Table 3.2 Notes/Domino field types. (continued)

Field Type	Workflow Considerations
Keyword field types	Several notes field types present the user with a list of choices. These include: Dialog List, Check Box, Radio Button, List Box, and Combo Box. The choices can be entered by the developer or derived from a formula. The Address dialog, Access Control List, and View dialog, as a means of selecting among choices, do not work with a browser.
Authors	Authors fields work in conjunction with Author access in the database access control list (ACL). If you assign a user Author access in the ACL, the user can read documents in the database but cannot edit even their own documents. Listing users in an Authors field expands access rights by allowing the listed users to edit documents they create. Entries in an Authors field cannot override the database access control list; they can only refine it. Users who have been assigned No Access to a database can never edit a document, even if you list them in an Authors field. Users who already have Editor (or higher) access to the database are not affected by an Authors field. Authors fields affect only users who have Author access to the database.
Readers	A Readers field explicitly lists the users who can read documents created from the form. This field type is an excellent method of managing documents through the workflow lifecycle. By the programmatic update of this field on State transitions, the ownership of a document can be established. The field value can be edited, computed or computed for display. Entries in a Readers field cannot override the database access control list; they can only refine it. Users who have been assigned *No Access* to a database can never read a document, even if you list them in a Readers field.

3.2 Replication

Replication refers to a Notes/Domino database's ability to synchronize multiple copies of itself throughout a network. Replication is a powerful feature for large-scale workflow applications. For performance considerations, multiple replicas of the same database may need to reside throughout the country or, for international processes, the world. Replication keeps data synchronized, so the integrity of the workflow cycle remains intact. Additionally, replication formulas can be utilized to selectively replicate subsets of data.

3.3 Programmability

The Notes/Domino environment offers a wide range of choices for development tools.

Starting with the venerable @Formula language, through LotusScript, JavaScript, and Java, there is a programming tool that is most appropriate for the task being implemented.

While many robust workflow applications have been built with only the formula language, the addition of LotusScript with Notes R4 brought Notes into the big leagues of application development. The emergence of the browser interface called for a mechanism to perform processing at the client, which JavaScript provides in R5. Only Java could fill the need for a lower level object-oriented language that could be used in the cross platform world that Notes has supported for some time. Together, these languages provide a toolkit enabling novice to intermediate level developers to construct basic workflow applications, such as we review later in this chapter. This toolkit also enables experienced developers to architect and build sophisticated workflow management functionality, as we shall see in subsequent chapters.

3.3.1 Agents

Agents are stand-alone programs that usually perform a specific task. Flexibility is one of an agent's strongest characteristics. Agents can be run in the foreground by users or in the background by servers, either on demand or on a scheduled basis. Agents can run on one or more servers, on a Notes client, or on the Web. Agents can be personal to a specific user or shared by many users. Finally, agents can be written in any of the Notes/Domino development languages, except JavaScript, and they may call other agents.

With all of that power and flexibility, some control mechanism and protection for the Notes infrastructure is certainly necessary. That need is filled, for the most part, by the Agent Manager task that supports the running and troubleshooting of agents. There is no direct user interface to the Agent Manger. It is a server task that performs its responsibilities, such as agent execution security, from parameters set in the server document and individual databases.

3.3.2 Actions

Actions may be comprised of formula language statements, pre-built actions, JavaScript, or LotusScript. They exist within a view or form design, and offer an efficient method of presenting routine tasks to the user.

3.3.3 Events

The programming capabilities with an application occur mostly in response to events in objects. The objects in Domino include those depicted in Figure 3.1. The events occurring in those objects, and the timing of those events, are presented in Table 3.3.

Table 3.3 Events occurring in Domino objects.

Object	Event Name	Timing
Database		
	Postopen (LS, F)	After a database is opened
	Postdocumentdelete (LS, F)	After a document is deleted (the document is still available)
	Queryclose (LS, F)	When a database is being closed
	Querydocumentdelete (LS, F)	Before a document is marked for deletion
	Querydocumentundelete (LS, F)	Before a document is unmarked for deletion
	Querydragdrop (LS, F)	Before a drag-and-drop operation in a view
	Postdragdrop (LS, F)	After a drag-and-drop operation in a view
	Initialize (LS)	When a database is being loaded
	Terminate (LS)	When a database is being closed
View or Folder		
	Queryopen (LS, F)	Before a view or folder is opened
	Postopen (LS, F)	After a view or folder is opened
	Regiondoubleclick (LS, F)	When a region in a calendar view is double-clicked
	Queryopendocument (LS, F)	Before a document is loaded
	Queryrecalc (LS, F)	Before a view or folder is refreshed
	Queryaddtofolder (LS, F)	Before a document is added to a folder

Table 3.3 Events occurring in Domino objects. (continued)

Object	Event Name	Timing
	Querypaste (LS, F)	Before a document is pasted
	Postpaste (LS, F)	After a paste operation
	Querydragdrop (LS, F)	Before a drag-and-drop operation in a calendar view
	Postdragdrop (LS, F)	After a drag-and-drop operation in a calendar view
	Queryclose (LS, F)	When a view or folder is being closed
	Initialize (LS)	When a view or folder is loaded
	Terminate (LS)	When a view or folder is being closed
Form or Subform		
	WebQueryOpen (F) — form only	Before a Web document displays†
	WebQuerySave (F) — form only	Before a Web document is saved†
Page		
	onHelpRequest (F)	When help is selected
	JS Header (JS)	When a document is being loaded
	onBlur (JS)	When an object is deselected
	onClick (JS)	When an object is selected
	onDblClick (JS)	When an object is selected with a double-click
	onFocus (JS)	When an editable field is selected
	onHelp (JS)	When help is selected
	onKeyDown (JS)	When a key is pressed
	onKeyPress (JS)	When a key is pressed
	onKeyUp (JS)	When a key is released
	onLoad (JS)	After a document is opened
	onMouseDown (JS)	When a mouse button is pressed
	onMouseMove (JS)	When the mouse is moved
	onMouseOut (JS)	When the mouse is moved out of an object
	onMouseOver (JS)	When the mouse is moved over an object
	onMouseUp (JS)	When a mouse button is released
	onReset (JS) — form only	When a document is reset

Table 3.3 Events occurring in Domino objects. (continued)

Object	Event Name	Timing
	onSubmit (JS) — form only	Before a document is submitted
	onUnload (JS)	Before a document is closed
	Queryopen (LS, F)	Before a document is opened
	Postopen (LS, F)	After a document is opened
	Querymodechange (LS, F)	Before a document is changed to read or edit mode
	Postmodechange (LS, F)	After a document is changed to read or edit mode
	Postrecalc (LS, F)	After a document is refreshed (and values are recalculated)
	Querysave (LS, F)	Before a document is saved
	Postsave (LS, F)	After a document is saved
	Queryclose (LS, F)	Before a document is closed
	Initialize (LS)	When a document is being loaded
	Terminate (LS)	After the document is closed
	Click (LS) — form only	When an object is selected
Field		
	onClick (JS)	When an object is selected
	onChange (JS)	When an object is changed
	onBlur (JS)	When an editable field is deselected
	onFocus (JS)	When an editable field is selected
	Entering (LS)	When an editable field is selected
	Exiting (LS)	When an editable field is deselected
	Initialize (LS)	When a document is being loaded (after the Form Initialize event)
	Terminate (LS)	When a document is being closed
Action, Button, or Hotspot		
	onBlur (JS)	When an object is deselected
	onClick (JS)	When an object is selected
	onDblClick (JS)	When an object is selected with a double-click
	onFocus (JS)	When an editable field is selected
	onHelp (JS)	When help is selected

Table 3.3 Events occurring in Domino objects. (continued)

Object	Event Name	Timing
	onKeyDown (JS)	When a key is pressed down
	onKeyUp (JS)	When a key is released
	onMouseDown (JS)	When a mouse button is pressed down
	onMouseMove (JS)	When the mouse is moved
	onMouseOut (JS)	When the mouse is moved out of an object
	onMouseOver (JS)	When the mouse is moved over an object
	onMouseUp (JS)	When a mouse button is released
	Click (LS, F)	When an object is selected
	Objectexecute (LS)	When an object is activated by an OLE2 server that is FX/NotesFlow™ enabled
	Initialize (LS)	When an object is being loaded
	Terminate (LS)	When an object is being closed
Agent		
	Action (F)	When an agent is run
	JavaAgent (J)	When an agent is run
	Initialize (LS)	When an agent is being loaded
	Terminate (LS)	When an agent is being closed
† WebQueryOpen and WebQuerySave must be a formula with the following syntax: @Command([ToolsRunMacro];"agentname").		

3.4 Messaging

Messaging, an integration chore for many workflow management tools, is core functionality in Notes/Domino. The mail database file can be used for traditional notification messages or as a customized work queue. The message capability is extended to the standard mail database file. This file also contains calendaring, scheduling, and to-do list functionality. These features provide a great deal of flexibility for a highly customized work management interface.

3.5 Calendaring and Scheduling

As mentioned in the Messaging section, the Notes mail database file has integrated calendaring and scheduling functionality. This design provides an individual work queue that can prioritize tasks, categorize them by status, and access tasks programmatically. A work queue of this design can be implemented for individuals, work groups, departments, or any nontraditional Web-based organization.

3.6 Incorporating Other Data Sources

Integrating data from non-Notes data sources will undoubtedly be a requirement in some workflow scenarios. Again, we are presented with a number of methods to access the data from Domino, which we can select based on the specifics of the application need.

Table 3.4 lists some of the choices for non-Notes data access and integration APIs and products.

Table 3.4 Connectivity options to non-Notes data.

Connectivity Solutions	Description
DECS (Domino Enterprise Connection Services)	A forms-based development tool that provides live access to enterprise data and applications, including relational databases, transaction systems, and Enterprise Resource Planning (ERP) systems.
LS:DO (LotusScript Data Object)	LotusScript access to any ODBC-compliant data sources.
JDBC (Java Database Connectivity)	Access from Java agents to relational data via Standard JDBC classes. A JDBC to ODBC bridge also ships with Domino.
Lotus Domino Connector LotusScript Classes	A unified object model with a consistent interface to programmatically access enterprise data and applications. These classes can be used with LotusScript or Java.
Lotus Domino Connectors	Modules that provide native connectivity to enterprise data sources. These connectors can be accessed through Domino Enterprise Connection Services or programmatically through Domino classes.

Table 3.4 Connectivity options to non-Notes data. (continued)

Connectivity Solutions	Description
	Connectors to DB2, Oracle, Sybase, text- and file-based systems, EDA/SQL, and ODBC are provided with the Domino server. Premium connectors to ERP applications, Transaction Monitors, and Directory systems are available separately. Note that the NotesSQL driver for ODBC access to Domino data is available free from the Lotus Web site.
LSX (LotusScript extensions)	Create custom objects that work natively with Domino applications as well as Java and OLE. Examples of LSX are MQSeries, SAP, DB2, and rich text. The DB2 LSX ships with Domino Release 5. These are classes for programming directly to the DB2 client access library.
Lotus Enterprise Integrator (Lotus Notes Pump)	Data distribution server that provides support for event-driven or scheduled high-volume transfer and data source synchronization.
Domino Connector Toolkit	Provide developers with tools and information to build additional Domino Connectors and classes available in Java or LotusScript.

3.7 The Document Library Template

One of the standard example templates that ships with Notes/Domino incorporates workflow functionality. This template, Document Library, serves as a good first look at the implementation of many of the components of workflow that we discussed in Chapter 2.

A document library application is an electronic filing cabinet that stores documents for access by the originator, other individuals, or a workgroup. The database might contain any kind of document that needs to be reviewed by a number of people or other workgroups. The application currently can be accessed from either a Web browser or Notes client. The document library has a review cycle and document archiving functions.

The application is fairly simplistic in its design and workflow functionality. It utilizes three forms:

- Document,
- Response, and
- Response to Response.

Those are both the names and document types of the forms. When you create a new document, you specify a subject and select a category for the document. The current user is recorded as the originator of the document.

After creating a document, you specify the review cycle options. First, you select the individuals that will review the document. Reviewers can be typed in as individual names, multiple names separated by commas, or can be selected from the Domino directory via a dialog box.

The second step is to specify the review options. The first option, "Type of review," offers the choices of "One reviewer as a time" (serial) or "All reviewers simultaneously" (parallel). The next field, "Time Limit Options", lets you indicate whether a review should have a time limit, and what will happen if the time limit is exceeded.

The choices for this field are:

- No time limit for each review,

- Move to next reviewer after time limit expires, or

- Keep sending reminders after time limit expires.

If you select either the "Move to next reviewer" or the "Keep sending reminders" choices, the field "Time Limit (days)" will be displayed. This field lets you input the number of days (in whole numbers only) a reviewer has to act on your document.

The last review option is the "Notify originator after" field. This field sends a notification to the originator either after the final review or after each interim review.

The final field on the form contains the content of the document to be reviewed (Figure 3.3).

Figure 3.3 A document library main document.

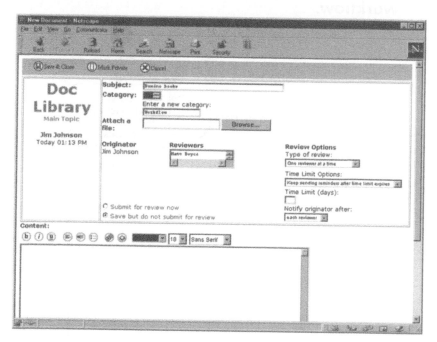

The workflow process implemented in document library (see Figure 3.4) has three states:

- New,
- In review, and
- Review complete.

The actions pertinent to the workflow are:

- save and close,
- submit for review, and
- all reviews complete.

Finally, the roles involved in this application are:

- Originator and
- Reviewer(s).

Figure 3.4 A state machine diagram for the document library workflow.

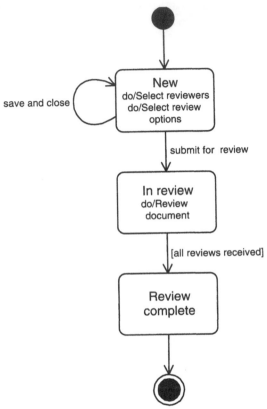

For Web usage, three agents are used: OpenDocument, SubmitDocument, and WebExpire.

The OpenDocument agent just resets a number of fields to their default values each time the document form is opened. These fields include: Save Options, Current User, Submit Now, Re-submit, and Web Categories. This agent is triggered by the WebQueryOpen event of the Document form.

The SubmitDocument agent is triggered by the WebQueryClose event of the Document form. This agent verifies that the document is being submitted for review. Then, it either sends notifications to all of the listed reviewers if the parallel review scenario has been selected, or to the next reviewer if the serial review scenario has been selected. Documents are then viewed by state, category, or author, as shown in Figure 3.5.

Figure 3.5 Documents listed by author.

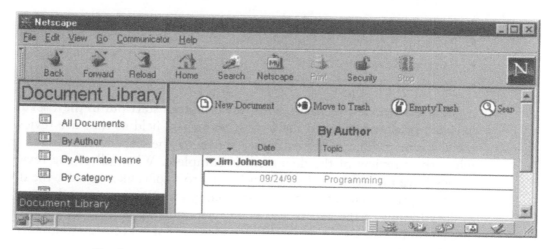

Reviewers may create a response document with their comments on the document under review, and mark their review as complete (see Figure 3.6).

Figure 3.6 Main document in review as seen by the first reviewer.

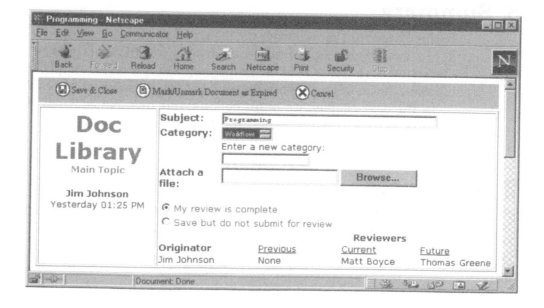

The reviewer management is performed through several "Computed for" display fields:

- Originator (`tmpOriginator`),
- Previous Reviewer (`tmpPrevReviewers`),
- Current Reviewer (`tmpCurrReviewer`), or
- Future Reviewer (`tmpFutReviewers`).

In a serial review scenario, all reviewer names start out in the `tmpFutReviewers` field and move to the `tmpCurrReviewer` field as they are reviewing the document. The reviewer names move to the `tmpPrevReviewers` field when their review of the document is complete. When all of the reviewer names have moved from `tmpFutReviewers` to `tmpPrevReviewers`, the document review is complete and the state changes to "Review Complete".

In addition to the agents and code contained in the various forms, this application also uses a script library (`SubmitForReview`) containing code specific to the workflow process.

For those new to workflow in Notes/Domino or new to workflow Web development, this application provides a quick study in simple workflow functionality.

3.8 Summary

In this chapter we moved from the theory of workflow presented in Chapter 2, to a basic review of Notes/Domino from a workflow perspective. We concluded with a review of a simple workflow-enabled application that employs the concepts of workflow and the functionality of Notes/Domino.

In the next chapter, we begin building our own Notes-based WFMS by starting with a common workflow task — notifying users of workflow actions and events.

3.9 References

1   Lotus Corporation. *HotMedia Connect for Domino*.
    [`http://www.lotus.com/home.nsf/welcome/hmc`]

4

Chapter 4

Notification

As we mentioned in Chapter 2, communication is an essential part of any workflow process. To carry out an action, you must know that there is something waiting to be done, or a document must be delivered to you in some fashion (e-mail, fax, letter, etc.). You must then carry out the actions required for the particular workflow state, and route the document to its next destination.

Assume that you are part of a workflow process in which you must approve a document. There are two types of notifications that may inform you that a document is waiting to be processed. The first type is based on a purely passive approach. That is, the document to be approved is sitting in a "basket" somewhere, waiting for you to pick it up. The basket might be a pile of forms deposited in a box, or the document may be sitting in a Notes database waiting for you to look at it through an "Unapproved documents" view. Either way, it is up to you to regularly check if anything is waiting for approval.

By contrast, a proactive notification is one where the document to be approved is delivered directly to you, or a message that you must process the document is delivered by some means. The form might be placed in the in-basket on top of your desk, received as an attachment to an e-mail, or you might be prompted to review it through a phone call or fax.

Documents delivered in a proactive manner demand attention in a way that passive notification cannot. It is too easy to ignore workflow processes that depend on the actors to periodically check for actions to complete. However, proactively notifying the actors in a process that something awaits their attention minimizes the chances that documents will be lost or trapped in some state awaiting the next action.

In this chapter, we create a proactive notification mechanism based on a Notes database. The database is developed as a reusable component that notifies a set of actors that actions are waiting to be completed as part of some workflow process. The proactive notifications supported are both hard-copy printouts and e-mail, using the Notes SMTP (Simple Mail Transport Protocol) transport system. Although this notification database can be used as a standalone component in any Notes-based system, it will be used in the remainder of this book as an essential component to our general workflow framework.

4.1 Modeling the Notification Process

We begin our design for the notification database by recognizing a key fact: proactive, automated notification is itself a workflow process. Thus, we can use the techniques outlined in Chapter 2 to model notifications. This model will guide the design of our database and the agents that carry out the notification tasks.

As in Chapter 2, start with a look at the "customer" and "vendor" involved in the notification process. As shown in Figure 4.1, the customer in this case is some workflow process that needs to notify one or more actors of actions to be performed. The vendor is our notification engine.

Figure 4.1 A communications-based model of proactive notification.

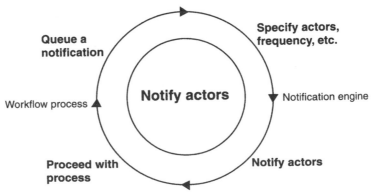

Queue a notification

Specify actors, frequency, etc.

Workflow process

Notify actors

Notification engine

Proceed with process

Notify actors

The workflow process itself initiates the request phase of the communications path, by queuing a request to perform a notification with the notification engine (i.e., our database). The negotiation phase is where the engine and process agree on the actors to be notified, the frequency of the notifications, the urgency, and when the notifications are no longer needed. The performance phase is carried out by the engine when it notifies the actors through whatever means it supports — in our case, this will be through e-mail. Finally, the acceptance phase is the continuing of the workflow process as the actors are notified and actions are carried out.

4.1.1 The Activity Model

Based on this model, we now diagram the activities that must be carried out to handle submission of requests from a process to our notification engine. In Figure 4.2, we have an activity diagram that shows the activities in their order of occurrence, from submission of different request types to processing of notifications by the engine.

Figure 4.2 An activity diagram of the notification process.

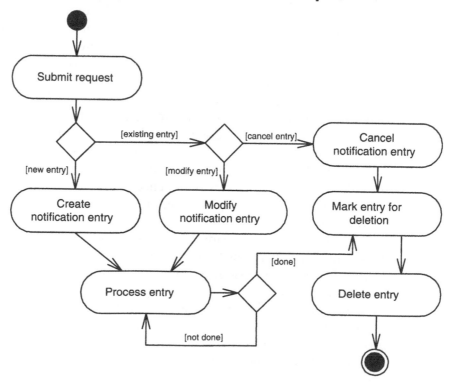

Note that upon submitting a request, two decisions are carried out. These decisions determine whether the request is a new notification entry, a modification of an existing entry, or a entry cancellation. This allows the workflow process that utilizes the notification engine to control both new and existing entries, in case a change in state or actor status requires the notification to change or be cancelled.

From the diagram, notice that cancelled requests cause the entry to be marked for deletion. Both new and modified entries are queued for processing by the engine. Once an entry is processed, another decision is made to determine whether further processing is required (i.e., repeat notification at some predefined time interval), or the notifications for that entry are complete.

Once notifications are complete, they are marked for deletion, and the "Delete entry" activity physically deletes the entries from the database. This also marks the end of the notification process for the submitted request.

4.1.2 The State Model

We have already said that our notification engine would be created using a Notes database. As the focus of Notes is on documents, and we see from the activity diagram that each notification document can be processed in a variety of ways, we should also model the document states. Figure 4.3 is a state diagram showing the various processing states through which a notification document progresses.

From this diagram, we can start to extract some Notes design elements that will help us create the notification database forms and views. For example, there are two types of documents that are "In process": new entries and modified entries. Thus, we should define two views: "In process\New" and "In process\Modified". Having modified documents in their own view will provide us with an easy way to check whether an incoming document already exists (note the guard conditions in Figure 4.3 for the New and Modified states).

Documents that are ready to be deleted will be in a "Cancelled" state. This defines another view that a Notes agent can use to automatically delete entries. Note that we also need some way to denote the view each document (i.e., notification entry) will be in. Thus, we will need a text field on the entry form to hold the current document state.

From the activity diagram, we see that one agent will be needed to pre-process incoming requests, one to process notification entries to generate e-mail, and another one to delete cancelled or completed entries.

Figure 4.3 A state diagram for notification documents.

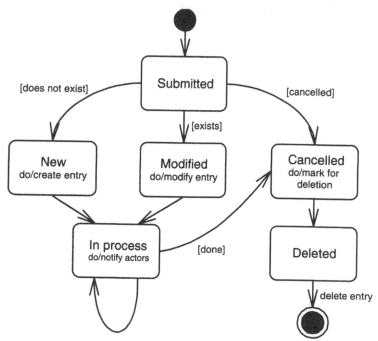

4.1.3 Evaluating the Model

Let's answer some questions about our model of the notification process. This will help us determine what features we need to think about when programming. Also, it will show where some features may not be supported. Table 4.1 summarizes our modeling Q&A evaluation.

Table 4.1 Evaluation of the notification workflow process.

Questions About the Model	Answers
What are the relevant subprocesses?	Creating new notification entries Modifying existing entries Canceling existing entries

Table 4.1 Evaluation of the notification workflow process. (continued)

Questions About the Model	Answers
What are the dependencies?	Database of notifications Notification entry form Agents to process notifications SMTP server process for sending mail
What triggers the process?	Request from another database or process
Who are the participants?	Requestor Notification database
What are the inputs?	Recipients, subject, text, notification schedule
What are the outputs?	E-mail messages to recipients
What are the business rules, implicit or explicit, that define, support, or limit the process?	Recipients must be listed in the server directory
What determines whether the process is a success?	E-mail messages received by designated recipients according to notification schedule
What exceptions or errors may occur?	Recipients may not have e-mail addresses

One dependency in our model is that recipients must be listed in the Domino directory. But what happens if the recipient is listed but does not have an e-mail address? We should take this into account in the database design. One way to accomplish that is to include a message to one or more recipients who are responsible for forwarding the notification to those recipients without registered e-mail addresses.

You may also note, from looking at the state diagram in Figure 4.3, that once an entry is cancelled, it may not be retrieved. This is a consequence of our model that we may or may not decide is important. In this case, we will leave the state diagram unchanged. In a real case, iterative design may be required to achieve consensus that the model is correct for the process it represents.

4.2 Prototyping the Database

The first step in developing the notification application is to create the database and define a notification entry form. You will find a copy of the database on the enclosed CD-ROM under the \ch04 subdirectory.

4.2.1 Designing the Form

We start by creating the database with a blank template in the R5 Domino Designer. Then, a new form titled "Notification Entry" is created. Figures 4.4–4.6 show the three parts of the entire notification entry form as it appears in the designer.

Figure 4.4 The notification entry form (top).

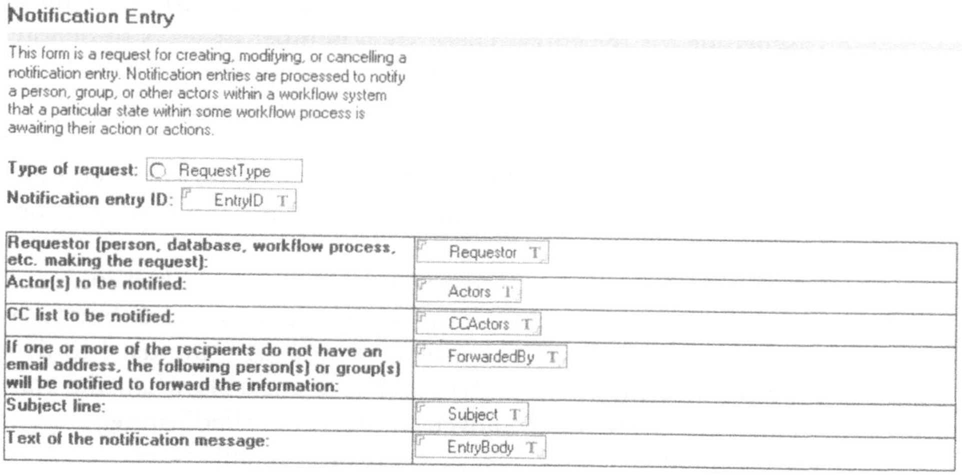

Notification Entry

This form is a request for creating, modifying, or cancelling a
notification entry. Notification entries are processed to notify
a person, group, or other actors within a workflow system
that a particular state within some workflow process is
awaiting their action or actions.

Type of request: ◯ RequestType
Notification entry ID: ⌐ EntryID T⌐

Requestor (person, database, workflow process, etc. making the request):	Requestor T
Actor(s) to be notified:	Actors T
CC list to be notified:	CCActors T
If one or more of the recipients do not have an email address, the following person(s) or group(s) will be notified to forward the information:	ForwardedBy T
Subject line:	Subject T
Text of the notification message:	EntryBody T

The top portion of the form (Figure 4.4) contains a *RequestType* radio button field with three values: "New", "Modification", and "Cancellation". *EntryID* stores the Notes document unique ID (UNID), computed with the formula @DocumentUniqueID.

The table contains six text fields. *Requestor* holds a string denoting who or what created the notification entry. *Actors* and *CCActors* are text lists of persons or groups defined in the Domino directory who are the recipients of the notification. The *ForwardedBy* field addresses the problem identified in the evaluation of our model. Those recipients without e-mail addresses listed in the Domino directory will be forwarded notifications by the persons or groups listed in this field. The remaining fields are the *Subject* and body (*EntryBody*) of the notification message, respectively.

Figure 4.5 The notification entry form (middle).

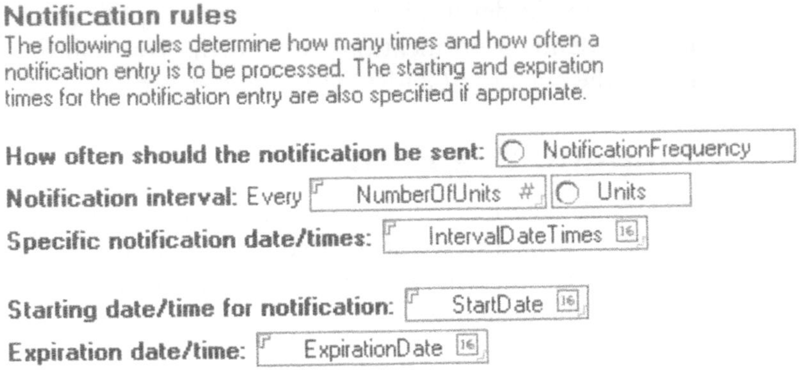

Notification rules
The following rules determine how many times and how often a
notification entry is to be processed. The starting and expiration
times for the notification entry are also specified if appropriate.

How often should the notification be sent: ⃝ NotificationFrequency

Notification interval: Every ⌐ NumberOfUnits # ⃝ Units

Specific notification date/times: ⌐ IntervalDateTimes 🗓

Starting date/time for notification: ⌐ StartDate 🗓

Expiration date/time: ⌐ ExpirationDate 🗓

The middle portion of the form (Figure 4.5) contains the notification schedule for the entry. There are a number of ways to denote a notification schedule, and this database design does not account for all of them. We have chosen a subset of methods for defining schedules.

The schedule is first specified by *NotificationFrequency*, a radio button that can take three values: "Once only", "At a specified interval", and "At specified dates/times". If the frequency is "Once only", the only required field to fill in is the *StartDate*.

If the frequency is selected to be "At a specified interval", the numeric field *NumberOfUnits* and the *Units* field (a radio button with values of "day(s)", "week(s)", "month(s)", or "N/A") are used to denote how often to process the entry. *ExpirationDate* controls when the state of the document is automatically moved to the "Cancelled" state.

The third method of specifying the schedule is to select specific date/times in the *IntervalDateTimes* field. Both of the latter schedule types require that the *ExpirationDate* field be filled in.

The bottom portion of the form (Figure 4.6) is used to hold precomputed values for specific fields. The fields are used to hold values that need to be precomputed only once for the recipients (*XMLActors* and *XMLCCActors*), the persons who will forward mail (*XMLForwardedBy*), and those recipients without e-mail addresses (*XMLOtherRecipients*). By precomputing these fields, the agent that processes each entry does not have to do Domino directory lookups or expand groups into the individual persons. Also, the fields will be formatted using eXtensible Markup Language, or XML[1]. This will facilitate parsing the information in the Java agent that processes the document to send the e-mail messages.

Figure 4.6 The notification entry form (bottom).

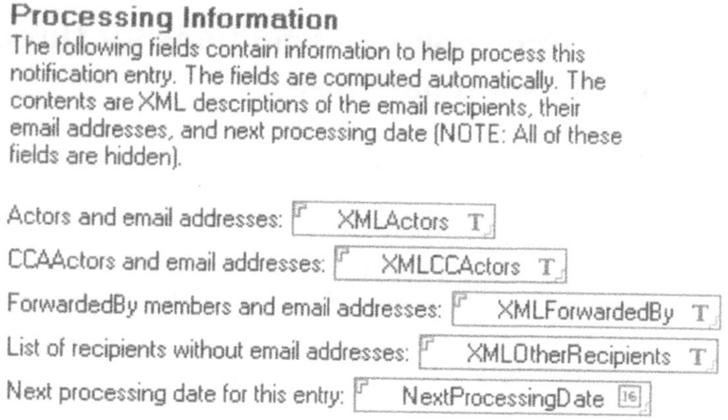

Processing Information

The following fields contain information to help process this notification entry. The fields are computed automatically. The contents are XML descriptions of the email recipients, their email addresses, and next processing date (NOTE: All of these fields are hidden).

Actors and email addresses: XMLActors   T

CCAActors and email addresses: XMLCCActors   T

ForwardedBy members and email addresses: XMLForwardedBy   T

List of recipients without email addresses: XMLOtherRecipients   T

Next processing date for this entry: NextProcessingDate 🗓

The last field, *NextProcessingDate*, determines when the document needs to be processed again. The Java agent can read the schedule information from the document to determine and set the next processing date.

4.2.2 Designing the Views

We need four basic views to facilitate the processing of the documents by agents. Three of the views, "Cancelled", "In process\New", and "In process\Modified", will use the same column definitions, but different view selection formula. The fourth view, "In process\All by date", provides the view for creating e-mail messages.

The "Cancelled", "In process\New", and "In process\Modified" views have four columns. The column definitions are provided in Table 4.2. The

"Notification frequency" column has a formula that takes the three different scheduling methods into account. It formats the output as in Figure 4.7.

Table 4.2 The common column definitions for three views in the notification database.

View Column	Formula
Start date	StartDate

Notification frequency	`u:=Units;`
	`t:=@If(u="1";"day(s)";u="2";"week(s)";`
	`u="3";"month(s)";"N/A");`
	`@If(NotificationFrequency="1";"Once only";`
	`NotificationFrequency="2";"Every " + @Text(NumberOfUnits) +`
	`" " + t;`
	`NotificationFrequency="3";@Text(IntervalDateTimes);"")`

Expiration date	ExpirationDate
Requestor	Requestor

Figure 4.7 The "Cancelled" view, showing the three different ways of formatting the notification schedule of a document.

	Start date	Notification frequency	Expiration date	Requestor
Workflow Notifier				
Cancelled	11/06/99	Every 2 week(s)	01/12/2000	Andrew Malley
In process	09/06/99	Once only	09/06/99	Robert Jones
All by date	08/24/99	08/25/99,09/25/99,10/10/99	10/10/99	Mary Smith
Modified				
New				

Table 4.3 shows the view selection formula for all four views. Note that all of the views are designed to show documents based on the "Notification Entry" form and based on the document status as set in the *RequestType* field (1 denoting "New", 2 denoting "Modification", and 3 denoting "Cancellation").

Table 4.3 The view selection formulas.

View Name	Selection Formula
Cancelled	`SELECT ((Form = "Notification Entry") &` `(RequestType="3"))`
In process\All by date	`SELECT ((Form = "Notification Entry") &` `((RequestType="1")	(RequestType="2")))`

Table 4.3 The view selection formulas. (continued)

View Name	Selection Formula
In process\Modified	`SELECT ((Form = "Notification Entry") &` `(RequestType="2"))`
In process\New	`SELECT ((Form = "Notification Entry") &` `(RequestType="1"))`

The last view, "In process\All by date", has the same four columns as the other views, but includes an additional first column. This column uses the formula `@Text(NextProcessingDate)`, categorized and sorted in descending order. The agent that processes the notifications will use this view to select the documents to be processed each day.

4.3 Writing the Database Agents

As we stated in developing our model, we need three agents to handle the notification entry documents. One agent will preprocess the documents to handle the lookup of the recipients in the Domino directory. Another agent will automatically delete documents. The third agent will generate e-mail notifications according to the document schedules.

In this section, we develop the code for all three agents. Note that we have carefully isolated the code into discrete tasks, with a limited number of tasks being done by a single agent. This avoids writing one overly complex agent that handles all of the tasks, exceptions, error conditions, etc. It also makes it easier to modify the design later, modifying or replacing selected agents rather than working on the whole design at once.

4.3.1 The Agent Structure

Note that all of the agents are written in Java, and we will write all of the Java agents in this book using the same general structure. You can see the class diagram illustrating the Java agent structure in Figure 4.8.

In the diagram, you see that the *myAgent* class (which we specialize in the following sections) is derived from the *lotus.domino.AgentBase* class. Inside *myAgent* is where we access the other *lotus.domino* classes: *Session*, *AgentContext*, and *Database*. Therefore, we include some static methods on the class (shown in the diagram as underlined methods) to provide convenient access to the instances of these Domino classes.

To provide for debugging and logging of the agent tasks, a *notesLogFile* class is defined. This class keeps a private instance of the *lotus.domino.Log* class to write messages to a log file and the Java console. *notesLogFile* is defined as a singleton[2] class, meaning that only one instance of the class is allowed to exist. A static *getInstance()* method returns the single instance of the class to whatever object needs to write to the log.

Each agent needs to perform one or more tasks. To keep the tasks separate from the base agent class, *myAgent* creates task objects. Each task object implements a *notesTask* interface that defines a *perform()* method. Thus, agents can do one or more tasks by simply creating the appropriate task object and calling its *perform()* method.

Figure 4.8 A class diagram depicting the general Java agent structure.

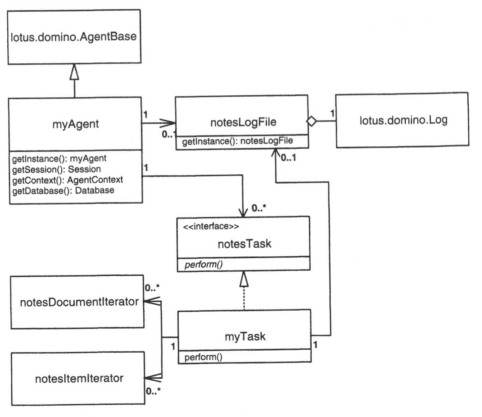

Finally, two helper classes are defined to ease the work involved in enumerating through documents or items in a *Database*, *DocumentCollection*, *View*, or *Document* object. These helper classes, *notesDocumentIterator* and *notesItemIterator*, manage the collections of documents and items. They provide an easy way to navigate through the objects while automatically cleaning up object memory through calls to the *refresh()* method of each object.

4.3.2 The *deleteEntries* Agent

Let's start with a simple agent that will delete the notification entry documents that have been completed or cancelled. The agent is set up as a new shared agent in the database, scheduled to run every week. Of course, you can adjust this schedule as desired through Domino Designer.

The entire code listings for the agent are shown in Listings 4.1–4.5. The first part of the agent, shown in Listing 4.1, is the code for the agent class that contains the *NotesMain()* method. There are four static variables in the class that point to instances of the agent itself, the Notes session, the agent context, and the current database. There are four static accessor methods that allow other classes to access these values, following our class structure diagram shown previously in Figure 4.8. We will see how these are used in the task implementation below.

Listing 4.1 The *deleteEntries* agent main class.

```
// **********************************************************************
//
// Module: deleteEntries.java
// Author: Dick Lam
//
// Description: This is a Lotus Notes Java agent that deletes all of the
//              documents in a specified view, using an instance of the
//              deleteEntriesTask class.
//
// **********************************************************************


import lotus.domino.*;


// **********************************************************************


public class deleteEntries extends AgentBase
{
```

Listing 4.1 The *deleteEntries* **agent main class. (continued)**

```
// NotesMain
public void NotesMain()
{
   try {
      // get the session and set up the log file
      myAgentInstance = this;
      mySession = getSession();
      notesLogFile.createInstance(mySession, "deleteEntries");

      // access the agent context and database
      myContext = mySession.getAgentContext();
      myDatabase = myContext.getCurrentDatabase();

      // carry out the agent tasks
      deleteEntriesTask task = new deleteEntriesTask("Cancelled");
      task.perform();
      task = null;

      // clean up
      myDatabase.recycle();
      myDatabase = null;
      myContext.recycle();
      myContext = null;
      notesLogFile.deleteInstance();
   }
   catch (Exception e) {
      notesLogFile.getInstance().write( "NotesMain exception: " +
                                        e.toString() );

   }
   finally {
      mySession = null;
   }
}
```

Listing 4.1 The *deleteEntries* **agent main class. (continued)**

```
// ------------------------------------------------------------------

// getInstance - returns the instance of the Notes agent
public static deleteEntries getInstance()
{
   return myAgentInstance;
}


// ------------------------------------------------------------------

// getSession - returns the Notes Session of the current agent
public static Session getAgentSession()
{
   return mySession;
}


// ------------------------------------------------------------------

// getContext - returns the Notes AgentContext of the current agent
public static AgentContext getContext()
{
   return myContext;
}


// ------------------------------------------------------------------

// getDatabase - returns the Notes Database of the current agent
public static Database getDatabase()
{
   return myDatabase;
}
```

Listing 4.1 The *deleteEntries* **agent main class. (continued)**

```
// ******************************************************************
// properties


// instance of this agent
private static deleteEntries myAgentInstance = null;


private static Session mySession = null;     // session, agent context,
private static AgentContext myContext = null;// and database for this agent
private static Database myDatabase = null;
}


// ******************************************************************


// end of deleteEntries.java
```

The *NotesMain()* method in Listing 4.1 has four main sections (separated by comments). The first two sections simply initialize the static variables and create an instance of the singleton class *notesLogFile*, using a call to the static method *notesLogFile.createInstance()*.

The third section of *NotesMain()* creates an instance of the task class *deleteEntriesTask*, and calls its *perform()* method. Once the task has been carried out, the fourth section of the code cleans up the Notes objects, deletes the *notesLogFile* instance, and exits.

Our second class, *notesLogFile*, is shown in Listing 4.2. It has a private constructor because the class is implemented as a singleton[1] (i.e., there is only one instance of the class allowed). The *createInstance()* and *deleteInstance()* methods are called once from within *NotesMain()* — other classes just retrieve the instance as needed through the *getInstance()* method.

Listing 4.2 The *notesLogFile* **class.**

```
// ******************************************************************
//
// Module: notesLogFile.java
// Author: Dick Lam
//
```

Listing 4.2 The *notesLogFile* **class. (continued)**

```java
// Description: This is a Java class that implements a singleton instance
//              of a Notes Log class for logging agent messages to an output
//              file and Java console.
//
// Usage: Once an agent has created a Session instance, call the static
//        method createInstance to create the singleton instance of the
//        notesLogFile. Other classes may then write to the log.
//
//        Session s = getSession();
//        notesLogFile.createInstance(s, "log");
//        notesLogFile.getInstance().write("This is a sample debug message.");
//        notesLogFile.deleteInstance();
//
// **************************************************************************

import java.io.*;
import lotus.domino.*;


// **************************************************************************

public class notesLogFile
{
   // getInstance - returns an instance of this class
   public static notesLogFile getInstance()
   {
      return singleInstance;
   }

   // -----------------------------------------------------------------------

   // createInstance - creates an instance of this class
   public static void createInstance(Session s, String logName)
   {
```

Listing 4.2 The *notesLogFile* **class. (continued)**

```
    try {
    // if there is no instance of this class, create one
    if (singleInstance == null) {
        singleInstance = new notesLogFile();

        singleInstance.myLog = s.createLog(logName);
        singleInstance.myLog.setOverwriteFile(true);
        singleInstance.myLog.openFileLog(logFile);
    }
  }
  catch (Exception e) {
     singleInstance = null;
  }
}


// ----------------------------------------------------------------------

// deleteInstance - deletes an instance of this class
public static void deleteInstance()
{
   try {
      // if there is an instance of this class, close and delete it
      if (singleInstance != null) {
         singleInstance.myLog.close();
         singleInstance.myLog = null;
         singleInstance = null;
      }
   }
   catch (Exception e) {
      singleInstance = null;
   }
}
```

Listing 4.2 The *notesLogFile* **class. (continued)**

```
// ---------------------------------------------------------------

// setLogFileName - sets the name of the log file to use (note: this
//                  method must be called before createInstance)
public static void setLogFileName(String fullPathName)
{
    logFile = new String(fullPathName);
}


// ---------------------------------------------------------------

// setDebugState - turns debugging log on/off
public static void setDebugState(boolean state)
{
    debug = state;
}


// ---------------------------------------------------------------

// setConsoleState - turns debugging Java console messages on/off
public static void setConsoleState(boolean state)
{
    mirror = state;
}


// ---------------------------------------------------------------

// write - writes a message to the log if it is active
public void write(String message)
{
    try {
        if ( (myLog != null) && debug ) {
            // write to the log file
```

Listing 4.2 The *notesLogFile* **class. (continued)**

```
            myLog.logAction(message);
            if (mirror) {
                // write to the Java console
                System.out.println(message);
            }
        }
    }
    catch (Exception e) {
        System.out.println( "notesLogFile.write() exception: " +
                            e.toString() );
    }
}


// -----------------------------------------------------------------


// constructor (private because this is a singleton)
private notesLogFile()
{
}


// *****************************************************************
// properties

// Notes log instance used by the singleton instance
private Log myLog = null;

// the singleton instance of this class
private static notesLogFile singleInstance = null;

// the log file name and flags for controlling output
private static boolean debug = true;      // to write messages to log file
private static boolean mirror = true;     // to also write to Java console
```

Listing 4.2 The *notesLogFile* **class. (continued)**

```
      private static String logFile = "c:\\temp\\agent.log";
}


// ********************************************************************


// end of notesLogFile.java
```

The method that creates the instance of this class actually initializes the private property of type *lotus.domino.Log*. This creates an output log file whose name defaults to c:\temp\agent.log. Note that agents may first call the static method *setLogFileName()* before creating the log file instance to change the pathname of the log file.

The other two static property setter methods turn the file log and the Java console log on or off. The final method, *write()*, places messages to the file log and Java console, depending on whether these are turned on or not.

Listing 4.3 contains the code for the third class, *notesTask*. This is the interface that user-defined task classes implement, defining a *perform()* method that the agent calls to carry out the task.

Listing 4.3 The *notesTask* **interface.**

```
// ********************************************************************
//
// Module: notesTask.java
// Author: Dick Lam
//
// Description: This is a Java interface that other classes implement to
//              perform some task for an agent.
//
// ********************************************************************


public interface notesTask
{
   // perform - performs this task
   public abstract void perform();
}
```

Listing 4.3 The *notesTask* interface. (continued)

```
// ******************************************************************


// end of notesTask.java
```

The actual task of deleting the entries in the "Cancelled" view is carried out by *deleteEntriesTask* (Listing 4.4). The constructor for this class accepts and stores the name of the view to be used for specifying the documents to be deleted. This view name is passed in by the agent when the task object is created (see Listing 4.1).

Listing 4.4 The *deleteEntriesTask* class.

```
// *******************************************************************
//
// Module: deleteEntriesTask.java
// Author: Dick Lam
//
// Description: This is a Java class that implements the notesTask interface.
//              The task of this class is to delete the documents in a view.
//
// *******************************************************************


import lotus.domino.*;


// *******************************************************************


public class deleteEntriesTask implements notesTask
{
   // constructor
   public deleteEntriesTask(String view)
   {
      // save the view name to process
      viewName = new String(view);
   }
```

Listing 4.4 The *deleteEntriesTask* **class. (continued)**

```
// ---------------------------------------------------------------

// perform - performs this task
public void perform()
{
   try {
      // make sure we have a view name
      if ( (viewName == null) || (viewName.length() == 0) )
         return;

      // get the view
      View v = deleteEntries.getInstance().getDatabase().getView(viewName);
      if (v == null)
         return;

      // iterate over the documents and remove them
      notesDocumentIterator i = new notesDocumentIterator(v);
      while ( i.hasMoreDocuments() ) {
         Document doc = i.nextDocument();
         if (doc != null)
            doc.remove(true);
      }

      // clean up
      i = null;
      v.clear();
      v = null;
   }
   catch (Exception e) {
      notesLogFile.getInstance().write( "deleteEntriesTask.perform(): " +
                                        e.toString() );
   }
}
```

Listing 4.4 The *deleteEntriesTask* **class. (continued)**

```
// ****************************************************************
// properties


    private String viewName = null;  // view to process
}


// ****************************************************************


                              \
// end of deleteEntriesTask.java
```

The *notesTask* interface method *perform()* carries out the deletion of the documents. In this method, you see why we defined the static accessor methods in the main agent class. The statement

```
View v = deleteEntries.getInstance().getDatabase().getView(viewName);
```

uses the static *getInstance()* method to retrieve the running instance of the main agent *deleteEntries*. The *getDatabase()* method on the agent object returns the current database, and *getView()* retrieves the view specified by the constructor of this task class.

Also, you see why the log class was defined as a singleton with a static method to retrieve the instance. This allows statements such as the following, used in processing any exceptions during the task.

```
notesLogFile.getInstance().write( "deleteEntriesTask.perform(): " +
    e.toString() );
```

We have avoided passing around all of the session, context, database, and log parameters by setting up our agent and helper classes.

If you examine the code fragment that enumerates through the documents of the view, you see the final helper class specified in our general structured agent design: *notesDocumentIterator*. The iterator object is generated using a constructor that takes the view as an argument. The iterator then behaves as a *java.util.Enumeration* type, with *hasMoreDocuments()* and *nextDocument()* allowing enumeration through the view to delete each document.

Our fifth and final listing for this agent, Listing 4.5, shows the code for *notesDocumentIterator*. This helper class uses four different constructors, enabling the iterator to work on a *Database*, a *DocumentCollection*, a *View*, or a *ViewNavigator* object.

Listing 4.5 The *notesDocumentIterator* **class.**

```
// **********************************************************************
//
// Module: notesDocumentIterator.java
// Author: Dick Lam
//
// Description: This is a Java class that facilitates iterating over a
//              document collection, a view, or the documents in a database.
//              This iterator is used to automatically recycle the memory
//              for loops over document collections.
//
// Usage: You can iterate over a document collection in a Database, a
//        DocumentCollection, a View, or a ViewNavigator. Construct the
//        iterator with the appropriate Notes object. For example, to iterate
//        over a set of documents in a View (not counting categories or view
//        totals):
//
//        notesDocumentIterator iter = new notesDocumentIterator(view);
//        while ( iter.hasMoreDocuments() ) {
//            Document doc = iter.nextDocument();
//            ...
//        }
//
// **********************************************************************

import lotus.domino.*;


// **********************************************************************
```

Listing 4.5 The *notesDocumentIterator* **class. (continued)**

```
public class notesDocumentIterator
{
   // constructor
   public notesDocumentIterator(Database db)
   {
      if (db != null) {
         collectionType = DATABASE;
         myDB = db;
         numberOfDocuments = countDocuments();
      }
   }


   // -----------------------------------------------------------------


   // constructor
   public notesDocumentIterator(DocumentCollection col)
   {
      if (col != null) {
         collectionType = COLLECTION;
         myDocs = col;
         numberOfDocuments = countDocuments();
      }
   }


   // -----------------------------------------------------------------


   // constructor
   public notesDocumentIterator(View v)
   {
      if (v != null) {
         collectionType = VIEW;
         myView = v;
         numberOfDocuments = countDocuments();
```

Listing 4.5 The *notesDocumentIterator* **class. (continued)**

```
      }
  }


  // --------------------------------------------------------------


  // constructor
  public notesDocumentIterator(ViewNavigator nav)
  {
     if (nav != null) {
        collectionType = VIEWNAVIGATOR;
        myViewNavigator = nav;
        numberOfDocuments = countDocuments();
     }
  }


  // --------------------------------------------------------------


  // hasMoreDocuments - returns true if there is another document
  public boolean hasMoreDocuments()
  {
     // if we have not started and there is at least one document,
     // or if there are documents remaining, return true
     if ( !started && (numberOfDocuments >= 1) )
        return true;
     else if (numberOfDocuments > 0)
        return true;

     try {
        // clean up
        if (curDoc != null) {
           curDoc.recycle();
           curDoc = null;
        }
```

Listing 4.5 The *notesDocumentIterator* **class. (continued)**

```
        if (curEntry != null) {
          curEntry.recycle();
          curEntry = null;

      }

    }

    catch (Exception e) {

    }


    return false;

  }


  // -----------------------------------------------------------------


  // nextDocument - returns the next document in the collection
  public Document nextDocument()
  {

    try {
      // make sure there is a document to return
      if (numberOfDocuments == 0)
        return null;


      // set the flag to indicate iteration has started
      if (!started) {
        started = true;
        curDoc = getFirst();
      } else {
        curDoc = getNext();
      }


      // decrement the number of documents remaining
      numberOfDocuments -= 1;
      return curDoc;

    }
```

Listing 4.5 The *notesDocumentIterator* **class. (continued)**

```
      catch (Exception e) {
        return null;
      }
    }


    // -----------------------------------------------------------------------


    // reset - resets iterator to the beginning
    public void reset()
    {
      try {
        started = false;
        numberOfDocuments = countDocuments();

        if (curDoc != null) {
          curDoc.recycle();
          curDoc = null;
        }
        if (lastDoc != null) {
          lastDoc.recycle();
          lastDoc = null;
        }
        if (curEntry != null) {
          curEntry.recycle();
          curEntry = null;
        }
        if (lastEntry != null) {
          lastEntry.recycle();
          lastEntry = null;
        }
      }
      catch (Exception e) {
      }
```

Listing 4.5 The *notesDocumentIterator* **class. (continued)**

```
}

// ------------------------------------------------------------------

// countDocuments - gets the total number of documents in the collection
private int countDocuments()
{
  try {
    switch (collectionType) {
      case DATABASE:
        myDocs = myDB.getAllDocuments();
        return myDocs.getCount();

      case COLLECTION:
        return myDocs.getCount();

      case VIEW:
        myViewEntryCollection = myView.getAllEntries();
        return myViewEntryCollection.getCount();

      case VIEWNAVIGATOR:
        return countDocuments(myViewNavigator);

      default:
        return 0;
    }
  }
  catch (Exception e) {
    return 0;
  }
}

// ------------------------------------------------------------------
```

Listing 4.5 The *notesDocumentIterator* **class. (continued)**

```
// countDocuments - gets the total number of documents in the collection
private int countDocuments(ViewNavigator nav)
{
   try {
      // find out how many documents this view navigator holds
      int count = 0;
      if (nav.getFirst() == null)
         return 0;
      else
         count = 1;

      while (nav.getNext() != null)
         count += 1;

      return count;
   }
   catch (Exception e) {
      return 0;
   }
}


// ---------------------------------------------------------------------------

// getFirst - gets the first document
private Document getFirst()
{
   try {
      switch (collectionType) {
         case DATABASE:
         case COLLECTION:
            return myDocs.getFirstDocument();
```

Listing 4.5 The *notesDocumentIterator* **class. (continued)**

```
            case VIEW:
                curEntry = myViewEntryCollection.getFirstEntry();
                return curEntry.getDocument();


            case VIEWNAVIGATOR:
                curEntry = myViewNavigator.getFirst();
                return curEntry.getDocument();


            default:
                return null;
        }
    }
    catch (Exception e) {
        return null;
    }
}


// ----------------------------------------------------------------------


// getNext - gets the next document
private Document getNext()
{
    try {
        switch (collectionType) {
            case DATABASE:
            case COLLECTION:
                lastDoc = curDoc;
                curDoc = myDocs.getNextDocument();
                lastDoc.recycle();
                lastDoc = null;
                return curDoc;
```

Listing 4.5 The *notesDocumentIterator* **class. (continued)**

```
            case VIEW:
                lastEntry = curEntry;
                curEntry = myViewEntryCollection.getNextEntry();
                lastEntry.recycle();
                lastEntry = null;
                return curEntry.getDocument();

            case VIEWNAVIGATOR:
                lastEntry = curEntry;
                curEntry = myViewNavigator.getNext();
                lastEntry.recycle();
                lastEntry = null:
                return curEntry.getDocument();

            default:
                return null;
        }
    }
    catch (Exception e) {
        return null:
    }
}


// *************************************************************************
// properties

// object to iterate over
private int collectionType;        // one of the constants defined below

private Database myDB = null;     // collections supported
private DocumentCollection myDocs = null;
private View myView = null;
private ViewEntryCollection myViewEntryCollection = null;
```

Listing 4.5 The *notesDocumentIterator* **class. (continued)**

```
        private ViewNavigator myViewNavigator = null;


        // variables to control iteration
        private boolean started = false;
        private int numberOfDocuments = 0;

        private Document curDoc = null;
        private Document lastDoc = null;
        private ViewEntry curEntry = null;
        private ViewEntry lastEntry = null;


        // type of collection
        private static final int DATABASE = 1;
        private static final int COLLECTION = 2;
        private static final int VIEW = 3;
        private static final int VIEWNAVIGATOR = 4;
}


// ***********************************************************************

// end of notesDocumentIterator.java
```

The private methods of this class handle the work of counting the number of documents in the relevant collection, iterating through them, and cleaning up memory through the *recycle()* method. A public *reset()* method also allows starting an iteration over from the beginning without creating a new iterator instance.

Now you might say to yourself (or maybe you already have), "Why all of this code to do such a simple function — delete some documents selected by a view?" The answer is, of course, that you do not *need* all that code for this one task. But, in the examples that follow in the remainder of the book, we will achieve a great deal of reuse from this code. Reuse is vitally important to rapidly prototype and deploy applications, using code that is already written, debugged, and documented.

4.3.3 The *prepareEntriesForProcessing* **Agent**

Before we can process any documents to generate e-mail, each one must be preprocessed to fill in the fields in the bottom portion of our "Notification Entry" form (i.e., the function of the next agent). Listing 4.6 is the main code for the agent. You will notice a great deal of similarity to the *Notes-Main()* method in the last agent.

Listing 4.6 The *prepareEntriesForProcessing* **agent class.**

```
// *************************************************************************
//
// Module: prepareEntriesForProcessing.java
// Author: Dick Lam
//
// Description: This is a Lotus Notes Java agent that prepares notification
//              entries for processing. The preparation involves lookup of
//              all email recipients to set fields in new or modified
//              notification entry documents. This limits the lookup of
//              email addresses to once per new or modified entry.
//
// *************************************************************************


import lotus.domino.*;


// *************************************************************************


public class prepareEntriesForProcessing extends AgentBase
{
    // NotesMain
    public void NotesMain()
    {
        try {
            // get the session and set up the log file
            myAgentInstance = this;
            mySession = getSession();
            notesLogFile.createInstance(mySession, "prepareEntriesForProcessing");
```

Listing 4.6 The *prepareEntriesForProcessing* **agent class. (continued)**

```
        // access the agent context and database
        myContext = mySession.getAgentContext();
        myDatabase = myContext.getCurrentDatabase();

        // carry out the agent tasks
        preprocessTask task = new preprocessTask();
        task.perform();
        task = null;

        // clean up
        myDatabase.recycle();
        myDatabase = null;
        myContext.recycle();
        myContext = null;
        notesLogFile.deleteInstance();
    }
    catch (Exception e) {
        notesLogFile.getInstance().write( "NotesMain exception: " +
                                          e.toString() );

    }
    finally {
        mySession = null;

    }
}


// ----------------------------------------------------------------------

// getInstance - returns the instance of the Notes agent
public static prepareEntriesForProcessing getInstance()
{
    return myAgentInstance;
}
```

Listing 4.6 The *prepareEntriesForProcessing* **agent class. (continued)**

```java
// ------------------------------------------------------------


// getSession - returns the Notes Session of the current agent
public static Session getAgentSession()
{
    return mySession;
}


// ------------------------------------------------------------


// getContext - returns the Notes AgentContext of the current agent
public static AgentContext getContext()
{
    return myContext;
}


// ------------------------------------------------------------


// getDatabase - returns the Notes Database of the current agent
public static Database getDatabase()
{
    return myDatabase;
}


// ****************************************************************
// properties

// instance of this agent
private static prepareEntriesForProcessing myAgentInstance = null;
```

Listing 4.6 The *prepareEntriesForProcessing* **agent class. (continued)**

```
        private static Session mySession = null;     // session, agent context,
        private static AgentContext myContext = null;// and database for this agent
        private static Database myDatabase = null;

}


// *****************************************************************************


// end of prepareEntriesForProcessing.java
```

The major differences are the change in data type for the *myAgentInstance* static variable (to match the class name), and the change in the class name of the task the agent invokes. The task class, *preprocessTask*, implements the *notesTask* interface, just as our previous task did.

The *perform()* method in our new task class must behave a little differently than our previous example. In this case, we want this agent to work only on new or modified documents. This prevents the agent from doing a lot of extra work, because the agent must carry out Domino directory lookups for all of the recipients of a notification entry to fill in the fields at the bottom of the form. Depending on the number of recipients, and whether or not some or all of the recipients are specified as groups in the directory, these lookup operations may take some time.

Listing 4.7 illustrates how to handle selection of new and modified documents only when processing documents.

Listing 4.7 The *perform()* **method of the** *preprocessTask* **class,**
illustrating the use of *AgentContext* **in processing only**
new or modified documents.

```
// perform - performs this task
public void perform()
{
   try {
      // get the documents that have not been processed
      AgentContext ac =
         prepareEntriesForProcessing.getInstance().getContext();
      DocumentCollection docs = ac.getUnprocessedDocuments();
```

Listing 4.7 The *perform()* **method of the** *preprocessTask* **class, illustrating the use of** *AgentContext* **in processing only new or modified documents. (continued)**

```
    // iterate over the documents to preprocess both the email
    // address information and the schedule
    notesDocumentIterator i = new notesDocumentIterator(docs);
    while ( i.hasMoreDocuments() ) {
        Document doc = i.nextDocument();
        if (doc != null) {
            preprocessAddresses(doc);

            // done with this document
            doc.save(true);
            ac.updateProcessedDoc(doc);
        }
    }

    // clean up
    i = null;
    ac = null;
    }
    catch (Exception e) {
        notesLogFile.getInstance().write( "preprocessTask.perform(): " +
                                            e.toString() );
    }
}
```

The key selection mechanism in this code is the call to the *AgentContext* method *getUnprocessedDocuments()*. This call returns a collection of documents that can be passed to an instance of our iterator class (reuse, again!) so that each document is processed. Once each document has been processed, another call to the *AgentContext* method *updateProcessedDoc()* marks this document so the agent will not process it again unless the document is modified.

The next part of our agent is the method called during the iteration through the document collection. This method, *preprocessAddresses()*

(Listing 4.8), handles the overall lookup strategy to determine the e-mail addresses for the recipients.

Listing 4.8 The *preprocessAddresses()* method to handle lookup and storage of the recipient e-mail addresses.

```
// preprocessAddresses - handles the recipient email addresses
private void preprocessAddresses(Document doc)
{
   try {
      // create the object to lookup email addresses in the Domino
      // directories
      directory = new dominoDirectoryLookup(
         prepareEntriesForProcessing.getInstance().getSession() );

      // create Hashtables to store the email addresses for each of the
      // computed fields, eliminating duplicates
      sendto = new Hashtable();
      cclist = new Hashtable();
      forward = new Hashtable();
      noemail = new Hashtable();

      // handle each address field
      scan(doc.getItemValue(Actors), sendto);
      scan(doc.getItemValue(CCActors), cclist);
      scan(doc.getItemValue(ForwardedBy), forward);
      storeAddresses(doc);    // store results in the 4 recipient fields

      // clean up
      sendto = null;
      cclist = null;
      forward = null;
      noemail = null;
      directory = null;
   }
```

Listing 4.8 The *preprocessAddresses()* **method to handle lookup and storage of the recipient e-mail addresses. (continued)**

```
catch (Exception e) {
    sendto = null;
    cclist = null;
    forward = null;
    noemail = null;
    directory = null;
    notesLogFile.getInstance().write(
        "preprocessTask.preprocessAddresses(): " + e.toString() );
    }
}
```

There are, of course, a variety of strategies for doing the name lookups. The one chosen here facilitates elimination of duplicates from the recipient lists, and provides a way to identify those recipients without e-mail addresses registered in the Domino directory. Thus, a list of recipients without e-mail addresses can be forwarded to one or more recipients (in the *Forwarded By* field of the entry form) that are responsible for contacting the people in the list.

The method starts with creation of an instance of another new class, *dominoDirectoryLookup*. This object will actually look up the e-mail addresses in the Domino directory. We'll examine this class later.

The next part of this method is to create four *java.util.Hashtable* objects, to hold the names and e-mail addresses for the main recipients, the carbon copy list, those responsible for forwarding the message, and those recipients without e-mail addresses. We chose hash tables to store the recipient information because it is a collection that does not allow duplicate entries. Thus, any duplicates that arise because the names appear multiple times, either directly or through group membership, are automatically removed.

The next three statements call the *scan()* method, shown in Listing 4.9. These statements get the recipient lists from the three fields of the notification entry form, look up the e-mail addresses, and add the names and e-mail addresses as keys and values, respectively, to the appropriate hash table. Finally, the call to *storeAddresses()* formats the contents of each hash table as XML and stores the XML strings in the fields at the bottom of the entry form. Thus, the e-mail addresses are readily available for the next agent to use in sending out notifications on schedule.

Listing 4.9 The *scan( )* method.

```
// scan - reads the information from the appropriate address field,
//         storing the persons with valid email addresses in the
//         Hashtable argument, and persons without valid email addresses
//         in the noemail collection
private void scan(Vector recipients, Hashtable table)
{
    try {
        // process each recipient
        for (int i = 0; i < recipients.size(); i++) {
            String rname = (String)recipients.elementAt(i);
            if ( (rname != null) && (rname.length() > 0) ) {
                // see if this is a person or group
                int type = directory.lookup(rname);
                switch (type) {
                    case dominoDirectoryLookup.PERSON:
                        // see if the person has an email address
                        String email = directory.getEmailAddress();
                        if ( (email == null) || (email.length() == 0) )
                            noemail.put(rname, "");    // notify separately
                        else
                            table.put(rname, email);   // email directly
                        break;

                    case dominoDirectoryLookup.GROUP:
                        // process each member of the group
                        scan(directory.getGroupMembers(), table);
                        break;
```

Listing 4.9 The *scan( )* method. (continued)

```
                    case dominoDirectoryLookup.NOTFOUND:
                    // notify separately
                    noemail.put(rname, "");
                    break;
                }
            }
        }
    }
    catch (Exception e) {
        notesLogFile.getInstance().write( "preprocessTask.scan: " +
                                    e.toString() );
    }
}
```

As shown in Listing 4.9, *scan( )* processes the list of recipients extracted from the multivalued text fields used to specify the recipients of each notification entry. For each entry in the list (an instance of *java.util.Vector*), the *dominoDirectoryLookup* instance is called. This object returns the type of entry, which can be a person, a group, or an indication that the name was not found in the directory. Each entry type is defined as a public static final variable in the *dominoDirectoryLookup* class.

If the lookup returns *PERSON*, the directory object's *getEmailAddress( )* method is called to extract the e-mail address corresponding to the last look up. The result is stored in the appropriate hash table, depending on whether the person has an e-mail address or not. All lookups that return *NOTFOUND* cause the name to be stored in the *noemail* hash table.

For those cases where the look up name is determined to be a *GROUP*, the scan method is called recursively. The argument to the recursive call is the list of members in the group. This allows us to handle an indeterminate nesting level of persons and groups in the recipient lists.

The final two methods in the *preprocessTask* class are shown in Listing 4.10. These methods format the names and addresses in each hash table as XML strings. The data are then stored in the appropriate field on the Notes form. The XML format is simple, consisting of a *RECIPIENTS* tag that contains individual *RECIPIENT* tags. Each *RECIPIENT* has two child tags, *NAME*

and *EMAIL*, to hold the name-e-mail pairs. Here is an example of the XML
format for two recipients.

```
<?xml version="1.0" standalone="yes"?>
<RECIPIENTS>
   <RECIPIENT>
      <NAME>Dan Giblin</NAME>
      <EMAIL>dgiblin@us.ibm.com</EMAIL>
   </RECIPIENT>
   <RECIPIENT>
      <NAME>Dick Lam</NAME>
      <EMAIL>rblam@us.ibm.com</EMAIL>
   </RECIPIENT>
</RECIPIENTS>
```

**Listing 4.10 The *storeAddresses()* and *toXML()* methods that format
and store the XML data in the document.**

```
// storeAddresses - stores the email addresses from the Hashtable
//                  collections into the document fields
private void storeAddresses(Document doc)
{
   try {
      doc.replaceItemValue( XMLActors, toXML(sendto) );
      doc.replaceItemValue( XMLCCActors, toXML(cclist) );
      doc.replaceItemValue( XMLForwardedBy, toXML(forward) );
      doc.replaceItemValue( XMLOtherRecipients, toXML(noemail) );
   }
   catch (Exception e) {
      notesLogFile.getInstance().write(
         "preprocessTask.storeAddresses(): " + e.toString() );
   }
}
```

Listing 4.10 The *storeAddresses()* **and** *toXML()* **methods that format and store the XML data in the document. (continued)**

```java
// ---------------------------------------------------------------------


// toXML - reads data from a Hashtable and formats it as an XML string
private String toXML(Hashtable table)
{
    // create a buffer and set up the XML enclosing tag
    StringBuffer buf = new StringBuffer(
        "<?xml version=\"1.0\" standalone=\"yes\"?>");
    buf.append("<RECIPIENTS>\n");

    try {
        // enumerate the table entries
        Enumeration e = table.keys();
        while ( e.hasMoreElements() ) {
            // get the key and value
            String key = (String)e.nextElement();
            if ( (key != null) && (key.length() > 0) ) {
                String value = (String)table.get(key);

                // output the XML tags
                buf.append("    <NAME>").append(key).append("</NAME>\n");
                buf.append("    <EMAIL>").append(value).append("</EMAIL>\n");
            }
        }
    }
    catch (Exception e) {
        notesLogFile.getInstance().write( "preprocessTask.toXML(): " +
                                    e.toString() );
    }
```

Listing 4.10 The *storeAddresses()* **and** *toXML()* **methods that format and store the XML data in the document. (continued)**

```
// done!
buf.append("</RECIPIENTS>\n");


return new String(buf);
}
```

We could have dispensed with this complexity and simply stored a list of e-mail addresses in multivalued text fields. However, this representation is easily generated and parsed by other agents, and it maintains a list of both names and their corresponding e-mail addresses. This allows us to expand the capabilities of the database later as other agents are added.

The last piece of the puzzle for this agent is the *dominoDirectoryLookup* class, shown in Listing 4.11. The constructor for this class simply calls a method on a Notes session to retrieve the list of Domino directory databases. The *lookup()* method then iterates through this database list, accessing the "People" and "Groups" views of each database. A call to the internal *find()* method searches the views for the target name.

Listing 4.11 The *dominoDirectoryLookup* **class.**

```
// *****************************************************************
//
// Module: dominoDirectoryLookup.java
// Author: Dick Lam
//
// Description: This is a Java class used to look up email addresses for
//              persons or groups registered in the Domino directory.
//
// *****************************************************************


import java.util.*;
import lotus.domino.*;


// *****************************************************************
```

Listing 4.11 The *dominoDirectoryLookup* **class. (continued)**

```
public class dominoDirectoryLookup
{
   // constructor
   public dominoDirectoryLookup(Session s)
   {
      try {
         // get the Domino directory lists known to this session
         myNABs = s.getAddressBooks();
      }
      catch (Exception e) {
         notesLogFile.getInstance().write( "dominoDirectoryLookup: " +
                                          e.toString() );
      }
   }


   // -------------------------------------------------------------------

   // lookup - finds the person or group corresponding to the input name
   //          in any of the valid address books (returns PERSON, GROUP,
   //          or NOTFOUND)
   public int lookup(String name)
   {
      try {
         // trim spaces from the name
         String searchName = name.trim();

         // iterate through the address books to look for the name
         int i = 0;
         while ( i < myNABs.size() ) {
            // open the next directory database
            curDB = (Database)myNABs.elementAt(i);
            if ( !curDB.isOpen() )
               curDB.open();
```

Listing 4.11 The *dominoDirectoryLookup* **class. (continued)**

```
// open person and group views on the current database
personView = curDB.getView(PERSONVIEW);
groupView = curDB.getView(GROUPVIEW);

// see if the name is in a person or group view (this sets
// the parameters to return with subsequent
// calls to the other methods below)
int type = find(searchName);
if (type == dominoDirectoryLookup.PERSON)
    return dominoDirectoryLookup.PERSON;
else if (type == dominoDirectoryLookup.GROUP)
    return dominoDirectoryLookup.GROUP;

// not found - try the next directory
personView.recycle();
personView = null;
groupView.recycle();
groupView = null;
curDB = null;
i += 1;
}

// clean up
curDB = null;

return dominoDirectoryLookup.NOTFOUND;
}
```

Listing 4.11 The *dominoDirectoryLookup* **class. (continued)**

```
        catch (Exception e) {
           personView = null;
           groupView = null;
           curDB = null;
           notesLogFile.getInstance().write( "dominoDirectoryLookup.lookup: " +
                                          e.toString() );
           return dominoDirectoryLookup.NOTFOUND;
        }
     }


     // -------------------------------------------------------------------


     // getEmailAddress - returns the current email address set by a
     //                   previous lookup
     public String getEmailAddress()
     {
        return new String(email);
     }


     // -------------------------------------------------------------------


     // getGroupMembers - returns the current list of group members set by a
     //                   previous lookup
     public Vector getGroupMembers()
     {
        return memberList;
     }


     // -------------------------------------------------------------------
```

Listing 4.11 The *dominoDirectoryLookup* **class. (continued)**

```
// find - finds members in a Domino directory database
private int find(String name)
{
   try {
      // see if the member is a person
      curDoc = getPerson(name);
      if (curDoc != null) {
         // extract the email address
         email = curDoc.getItemValueString(EMAILFIELD);
         return dominoDirectoryLookup.PERSON;
      }


      // see if the member is a group name
      curDoc = getGroup(name);
      if (curDoc != null) {
         // extract information about each member of the group and
         // return the array of results
         memberList = curDoc.getItemValue(MEMBERLIST);
         return dominoDirectoryLookup.GROUP;
      }


      // not found
      return dominoDirectoryLookup.NOTFOUND;
   }
   catch (Exception e) {
      curDoc = null;
      email = null;
      memberList = null;
      notesLogFile.getInstance().write( "dominoDirectoryLookup.find: " +
                                          e.toString() );
      return dominoDirectoryLookup.NOTFOUND;
   }
}
```

Listing 4.11 The *dominoDirectoryLookup* **class. (continued)**

```
// ----------------------------------------------------------------

// getPerson - retrieves a Person document by name
private Document getPerson(String name)
{
   try {
      // form a query and reset the person view to search
      String queryOnNames = "FIELD FullName contains \"" + name + "\"";
      personView.clear();

      // search the person documents for a match
      int matches = personView.FTSearch(queryOnNames);
      if (matches == 0)
         return null;
      else
         return personView.getFirstDocument();
   }
   catch (Exception e) {
      notesLogFile.getInstance().write(
         "dominoDirectoryLookup.getPerson: " + e.toString() );
      return null;
   }
}
```

Listing 4.11 The *dominoDirectoryLookup* **class. (continued)**

```
// -----------------------------------------------------------------------

// getGroup - retrieves a Group document by name
private Document getGroup(String name)
{
  try {
    // form a query and reset the group view to search
    String queryOnGroups = "FIELD ListName contains \"" + name + "\"";
    groupView.clear();

    // search the group documents for a match
    int matches = groupView.FTSearch(queryOnGroups);
    if (matches == 0)
      return null;
    else
      return groupView.getFirstDocument();
  }
  catch (Exception e) {
    notesLogFile.getInstance().write(
      "dominoDirectoryLookup.getGroup: " + e.toString() );
    return null;
  }
}


// ********************************************************************
// properties

private Vector myNABs = null;        // list of Domino directory databases

private Database curDB = null;        // current database
private View personView = null;       // views to open
private View groupView = null;
```

Listing 4.11 The *dominoDirectoryLookup* class. (continued)

```
        private View curView = null;        // current view
        private Document curDoc = null;      // current document (person or group)
        private String email = null;         // current email address
        private Vector memberList = null;    // current group member list


        // view and field names to use
        private final String PERSONVIEW = "People";
        private final String GROUPVIEW = "Groups";
        private final String MEMBERLIST = "Members";
        private final String EMAILFIELD = "MailAddress";


        // return types from a call to the lookup method
        public static final int PERSON = 1;
        public static final int GROUP = 2;
        public static final int NOTFOUND = 3;
}


// **************************************************************************


// end of dominoDirectoryLookup.java
```

The internal search calls the *getPerson()* and *getGroup()* methods to carry out full text searches on the two views. The results are stored in private variables that are retrieved with accessor methods *getEmailAddress()* and *getGroupMembers()*. Ideally, we would have returned the information about the person or group directly in the *lookup()* method by defining and using another data type. We implemented the class this way for efficiency, based on the way we structured our preprocessing task. However, note that this class is not thread-safe, so it would not be appropriate for a multi-threaded agent where each thread was using the same instance of *dominoDirectoryLookup*.

4.3.4 The *processEntries* **Agent**

Finally, we are ready to write the agent that does the work of performing the notifications — that is, creating and delivering the e-mail messages on schedule to the designated recipients. We will not show the main agent class, as it is similar to our previous two agents. You can view the code in the \ch04 subdirectory of the CD-ROM.

The task implemented by the *processEntries* agent is called, appropriately enough, *processTask*. As this task concerns itself with selecting the documents to be processed, updating the next processing date, and sending mail messages, it needs to access the current date. This is done in the task constructor, as shown in the following code fragment.

```
// create a DateTime object with today's date
today = processEntries.getInstance().getSession().createDateTime("Today");
```

Note that we use our structured agent methods *getInstance()* and *get-Session()* to retrieve the current Notes session. The variable *today* is stored as a private instance variable of type *DateTime*.

In the *perform()* method of the task, we select the documents to process by today's date. To do this, we access the view "In process\All by date", and create a view navigator object using the following code. (Note that the code requires two back slashes to refer to the view in the "In process" folder.)

```
View v = processEntries.getInstance().getDatabase().getView(
        "In process\\All by date");
ViewNavigator nav = v.createViewNavFromCategory( today.getDateOnly() );
```

Remember that we defined this view to use the formula @Text(NextProcessingDate) as the categorized, sorted, first column definition. Thus, the view navigator points to only those documents whose *NextProcessingDate* field matches today's date. We then use the view navigator as an argument to the constructor of *notesDocumentIterator* to iterate through our selected documents.

For each document being processed, we call two methods: *sendNotices()* and *updateSchedule()* (see the complete class listing on the CD-ROM). Let's consider the first method, which calls three additional methods. The first of these is *getAddresses()*, which extracts the XML version of the e-mail addresses from the document, along with the subject and body of the message to send. The second method is *sendMainMessage()*, shown in Listing 4.12.

Listing 4.12 The *sendMainMessage()* **method of the** *processTask* **class.**

```
// sendMainMessage - sends the main message document
private void sendMainMessage()
{
  try {
    // create the document
    Document mainMessage =
       processEntries.getInstance().getDatabase().createDocument();
    mainMessage.appendItemValue("Form", "Memo");

    // set the fields in the document to the appropriate values
    mainMessage.appendItemValue( "SendTo", getEmail(actors) );
    mainMessage.appendItemValue( "CopyTo", getEmail(ccactors) );
    mainMessage.appendItemValue("Subject", msgSubject);
    mainMessage.appendItemValue("Body", msgBody);

    // send the message
    mainMessage.setSaveMessageOnSend(false);
    mainMessage.send( getEmail(actors + ccactors) );
  }
  catch (Exception e) {
    notesLogFile.getInstance().write(
       "processTask.sendMainMessage(): " + e.toString() );
  }
}
```

The message is constructed as a new document based on the Notes form "Memo". The *SendTo* and *CopyTo* fields of the mail message are set to their corresponding lists of e-mail addresses. The addresses are extracted from the XML strings using the *getEmail()* method (Listing 4.13). Finally, the message is sent (but not saved in the database) by calling the *setSaveMessageOnSend()* with an argument of `false`, and calling the *send()* method on the new document.

Listing 4.13 The *getEmail()* **method for extracting e-mail addresses from an XML string representation.**

```
// getEmail - extracts email addresses from an XML string and returns
//            them in a Vector
private Vector getEmail(String xml)
{
   try {
      // create a vector to return and search through the XML for the
      // start and end tags
      Vector v = new Vector();
      int curIndex = 0, startIndex, endIndex;
      boolean done = false;

      while (!done) {
         // find the next starting tag
         startIndex = xml.indexOf("<EMAIL>", curIndex);
         if (startIndex == -1)
            done = true;
         else {
            // find the matching end tag
            startIndex += 7;
            endIndex = xml.indexOf("</EMAIL>", startIndex);
            if (endIndex == -1)
               done = true;
            else {
               v.addElement( xml.substring(startIndex, endIndex) );
               curIndex = endIndex + 8;
            }
         }
      }
```

Listing 4.13 The *getEmail()* **method for extracting e-mail addresses from an XML string representation. (continued)**

```
    return v;
  }
  catch (Exception e) {
    notesLogFile.getInstance().write(
      "processTask.getEmail(): " + e.toString() );
    return new Vector();
  }
}
```

As Listing 4.13 shows, the e-mail address extraction from our XML representation is simple. It relies on searching for the *EMAIL* start and end tags using the *java.util.String indexOf()* method. If the XML were more complex, or you would wish to validate it against a Document Type Definition (DTD), check formatting, etc., you would use another parser based on a tokenizer object or a full-featured Java-based XML parser[3].

The third and last method called from *sendNotices()* is *sendForwardMessage()*. This method is available on the CD-ROM, and uses exactly the same structure as the previous methods. The major difference is that the message is sent to the recipients in the *ForwardedBy* field, and the message includes the list of names of recipients who do not have e-mail addresses.

To complete the processing of the document, we must set the value of the *NextProcessingDate* field so that the document will be re-selected on the appropriate day. We must also check if the document has reached its expiration date, and if so, change its state so it will be deleted by the *deleteEntries* agent. These tasks are done by *updateSchedule()*, which executes the following code fragment.

```
String freq = doc.getItemValueString(NotificationFrequency);
if ( freq.equals("1") )
  scheduleOnce(doc);
else if ( freq.equals("2") )
  scheduleIntervals(doc);
else if ( freq.equals("3") )
  scheduleSpecified(doc);
```

For notifications that are processed one time only, the *scheduleOnce()* method is called. Its only function is to change the status of the request to "Cancelled", using the following statement.

```
doc.replaceItemValue(RequestType, "3");
```

To handle rescheduling at a specified interval, the *scheduleIntervals()* method is called (Listing 4.14). This method first retrieves the number of units and the unit of time interval (days, weeks, or months) from the notification entry document. It then creates a new *DateTime* object representing the current day, and calls either *adjustDay* or *adjustMonth* to add the appropriate time interval to the date. Finally, it checks the new date against the expiration date of the document, and cancels the document if necessary.

Listing 4.14 The *scheduleIntervals()* **method of the** *processTask* **class.**

```
// scheduleIntervals - add the interval to this date
private void scheduleIntervals(Document doc)
{
    try {
        // get the interval
        int numUnits = doc.getItemValueInteger(NumberOfUnits);
        String units = doc.getItemValueString(Units);

        // create a DateTime object with today's date and add the interval
        DateTime next =
            processEntries.getInstance().getSession().createDateTime("Today");
        if ( units.equals("1") )
            next.adjustDay(numUnits);       // days
        else if ( units.equals("2") )
            next.adjustDay(7*numUnits);     // weeks
        else if ( units.equals("3") )
            next.adjustMonth(numUnits);     // months

        // set the next processing date
        setDate(doc, NextProcessingDate, next);
```

Listing 4.14 The *scheduleIntervals()* **method of the** *processTask*
class. (continued)

```
      // if the next date is after the expiration date, cancel this entry
      Date expire = getDate(doc, ExpirationDate).toJavaDate();
      Date test = next.toJavaDate();
      if ( test.after(expire) )
         doc.replaceItemValue(RequestType, "3");


      // clean up
      test = null;
      expire = null;
      next = null;
   }
   catch (Exception e) {
      notesLogFile.getInstance().write(
         "processTask.scheduleIntervals(): " + e.toString() );
   }
}
```

The last method, *scheduleSpecified()*, handles the case of notifications
on user-specified dates. In this case, the method gets a list of dates from the
document's *IntervalDateTimes* field, and compares them one by one to the
current date (Listing 4.15). The dates to be compared are converted to
java.util.Date objects using the *toJavaDate()* method of *DateTime* to
facilitate comparison using the *after()* method.

Listing 4.15 The *scheduleSpecified()* **method of the** *processTask*
class.

```
// scheduleSpecified - generate notices at specified date/times
private void scheduleSpecified(Document doc)
{
   try {
      // get the Java version of today's date
      Date jtoday = today.toJavaDate();
```

Listing 4.15 The *scheduleSpecified()* **method of the** *processTask* **class. (continued)**

```
// get the list of specified Date/Times
Vector v = doc.getItemValue(IntervalDateTimes);

// go through the list (assumed to be ordered) to find the
// next scheduled date
int i = 0;
boolean done = false, found = false;

while ( !done && ( i < v.size() ) ) {
    // get the next element
    DateTime next = (DateTime)v.elementAt(i++);
    Date jnext = next.toJavaDate();

    // see if this date is after today
    if ( jnext.after(jtoday) ) {
        // set the next processing date
        setDate(doc, NextProcessingDate, next);
        found = true;
        done = true;
    }

    // clean up
    jnext = null;
    next = null;
}

// if a new date was not found, cancel this entry
if (!found)
    doc.replaceItemValue(RequestType, "3");
```

Listing 4.15 The *scheduleSpecified()* **method of the** *processTask*
class. (continued)

```
        // clean up
        v = null;
        jtoday = null;
    }
    catch (Exception e) {
      notesLogFile.getInstance().write(
        "processTask.scheduleSpecified(): " + e.toString() );
    }
  }
```

Note that this method makes the assumption that the dates in the field
are in increasing order. If no date past the current date is found in the list,
the document is cancelled.

4.4 Using the Notification Database

The \ch04 subdirectory on the enclosed CD-ROM contains all of the agent
code listings and a copy of the notification database. To use the database,
install it on a Domino server that has SMTP enabled. Set up each agent's
schedule as desired, and create some notification documents to test the operation.

4.5 Summary

In this chapter, we developed a critical component of a generalized workflow framework. This component, a notification engine, is a reusable database that maintains notification information for any workflow process. We
developed Java-based agents to process the notification entry documents
and automatically generate e-mail to a list of recipients. In the process, we
ended up with a generalized structure for Java agents and their associated
tasks. The structure contains several reusable components that help simplify
the writing of future agents.

In the next chapter, we reuse our notification database in the context of a
Domino-based workflow management system. We develop the framework
and agents for defining actions, roles, states, and rules to use in implementing your own Internet or intranet workflow applications.

4.6 **References**

1   Elliotte Rusty Harold. 1998. *XML: Extensible Markup Language*. New York, NY: IDG Books Worldwide.

2   Erich Gamma, et.al. 1994. *Design Patterns: Elements of Reusable Object-Oriented Software*. Reading, MA: Addison-Wesley.

3   Hiroshi Maruyama, et.al. 1999. *XML and Java*. Reading, MA: Addison-Wesley.

Chapter 5

Building a Workflow Management System

In the first three chapters, you learned the basics of workflow system design. We explained UML diagramming techniques, Notes/Domino features critical to Web application development, a basic approach to modeling business processes, and the steps involved in creating workflow systems. Our last chapter used these techniques in the development of a key building block for workflow — notification.

What we want to accomplish in this chapter is the creation of a general Notes/Domino-based workflow management system. We recognize at the outset that building a complete, general-purpose workflow management system is a monumental task. The intent is to provide a framework you can reuse to create your own workflow applications. By illustrating all of the steps involved in creating the framework, you will understand both the power and limitations of the framework. Using this knowledge, you can use the framework as is, or you can extend its capabilities as needed to model your or your customers' business processes.

5.1 An Example Workflow Model

To get started, we need to pick an example business process to model. This provides a basis for designing the features we wish to include in our framework. For this purpose, let's choose a model familiar to most people in business — travel expense accounting.

Obviously, travel expense reimbursement processes vary widely from company to company. The example process modeled in Figure 5.1, while simplified, will serve our purpose as a typical workflow that the framework should support.

Figure 5.1 A travel expense accounting workflow state diagram.

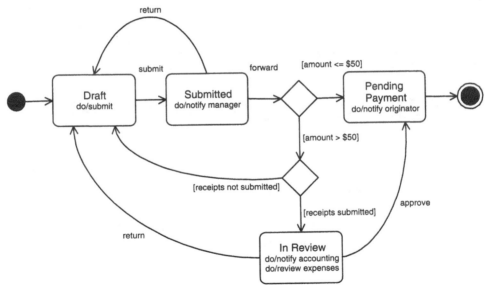

Note that there are several features illustrated in the state diagram. The "Draft" state has an action to submit a travel expense form for processing. An entry process in the "Submitted" state notifies the originator's manager that the submitted form must be approved. The manager role has two possible actions: forward the form for review, or return it to the originator for revision.

Two decision points illustrate some business rules that are coded into the workflow process. The first decision point is a test that compares the reimbursement amount with a limit (e.g., fifty dollars). If that limit is not exceeded, the form requires no further review and is moved directly to the "Pending Payment" state. Otherwise, another rule is applied. This rule checks whether or not expense receipts have been submitted. If not, the form is returned to the originator. Otherwise, the state is changed to "In Review".

The "In Review" state includes a notification to the accounting department. Accounting either returns the form to the originator for revision, or approves the document and changes the state to "Pending Payment". Finally, the "Pending Payment" state includes an entry action that notifies the originator that the document was approved.

5.2 Designing the Framework

Let's consider the components necessary for a usable workflow framework. As noted in Chapter 2, there are various models used to describe workflow. Here, we use the States/Actions/Roles model. To accommodate the need for encoding business rules in typical workflow processes (as shown in Figure 5.1), we add a fourth component: Rules.

It is worth diagramming the relationships among these four main objects to clarify how the framework will be built within the Notes/Domino environment. To do this, we first start with a schematic diagram depicting a Domino application (or database) with a specific form defined. Documents that will be processed by the workflow portion of the application are based on the form (see Figure 5.2).

Figure 5.2 A Domino application design schematic showing the relationship between forms and documents.

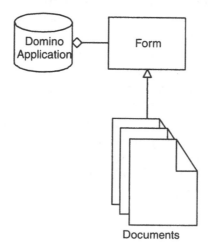

To process any document, it must be associated with a particular state. Thus, the document's form should define a state field that, depending on its value, determines the state of each document in the workflow process. Furthermore, the actions available for a particular document depend on both the state of the document and the role of the user who is currently viewing it. Thus, we need a package of design elements to encompass the definition of states, actions, and roles. This forms a Workflow Definition package.

Note that one of the goals for the framework is the ability to workflow-enable existing applications. We want to disturb as little of an existing application as possible. Thus, we place the state field into a subform, and specify this Workflow Subform as a component of the application's main form. This encapsulates the workflow functionality in a specific location within the target form design.

Our next design goal is to be able to launch specific tasks as states are entered and exited. The natural mechanism to accomplish this in Domino is with *WebQueryOpen* and *WebQuerySave* agents. These become part of our Workflow Agents package.

As shown in the diagram, we need to include business rule logic to associate rules (i.e., decision points) with particular actions. These rules allow the framework to make decisions within a particular process, and route documents to specific states based on these rule actions. This requires a Rule Engine package.

Finally, workflow processes must notify the relevant users that documents are in particular states, actions are required, etc. For this requirement, we include the notifier database and Notification Engine agent package developed in the last chapter. These elements are all diagrammed in Figure 5.3, showing how they mesh with the general Domino-based architecture of Figure 5.2.

Before we begin building each of the framework components, let's break down the States/Actions/Roles/Rules objects using a static structure diagram. The diagram is shown in Figure 5.4 (see page 124). It illustrates more detail about the relationships among the main objects.

Each state in a workflow process contains zero or more processing options. These options may include notifications to document originators, process administrators, state owners, etc. Options are usually selected in the workflow setup of each state, and may be initiated when the state is either entered or exited.

Figure 5.3 The Domino-WFMS framework architecture.

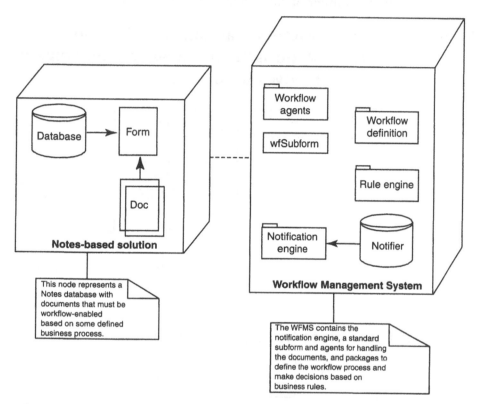

The next part of a workflow definition is the specification of the actions associated with each state. Note that each action is associated with one or more states, because there may be more than one previous state in which an action can be performed to transition to the same ending state (e.g., "Pending" to "Approval", or "Second Level Review" to "Approval").

Actions may include zero or more processing options, as well. For example, a processing option associated with an action might be to log the action to a file.

To implement the decision points in a state diagram, we use rules. Rules are contained within particular actions. We consider three general cases of state transition actions. First, leaving a state via a simple (i.e., non-rule-based) action can initiate direct state-to-state transitions (Figure 5.5A, see page 124). This is the most common type of state transition. Second, an action that contains a single rule can handle the transition from one state to either of two target states (Figure 5.5B). Third, if an action contains several rules (Figure 5.5C), any number of target states can be reached, depending on the number of rules defined for the action.

Figure 5.4 A static structure diagram showing the States/Actions/Roles/Rules relationships modeled by the workflow framework.

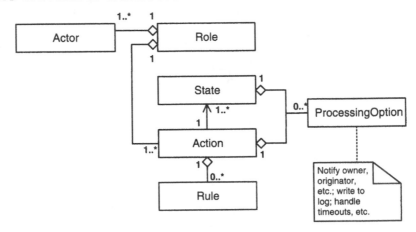

Figure 5.5 Three state transition action types.

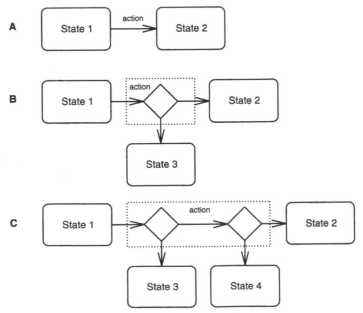

Finally, for each role in the workflow, there are at least one and perhaps more actors (individuals or groups) that act in that role. Each of these roles is associated with one or more actions. These relationships determine who can perform a particular action, and at what state they are allowed to perform the action.

As we will see after we completely build our framework, this set of relationships defines a specific order to defining our workflow process. The use of the framework, including the specification of actors, roles, actions, states, rules, and processing options, will be discussed in the last two chapters.

5.3 Building the Framework Forms

As previously stated, our Workflow Management System will be comprised of a Workflow Definition package, Workflow Agents, a Workflow Subform, a Rule Engine, and the Notification Engine from Chapter 4.

The Workflow Definition contains design elements that provide State, Action, and Role attributes to a workflow-enabled application. To build the Workflow Definition package, we must create a separate form for all three elements. See the sample database chap5.nsf on the CD-ROM for the complete design of all forms and views described in the following text.

5.3.1 The State Form

The wf_State form defines a workflow state by specifying a unique name and number for each state. For accountability and integrity of the workflow process, every state must have a state owner. This is a person or group that has ultimate responsibility for the document while it resides in this state. That individual owner or group is identified in our wf_State form.

A document resides in a state for a finite period of time, unless it is the final state in a workflow process, so we assign a duration period to each of our states. It could be a day, a week, or a year, but having a time period for each state will expedite the overall business process and preclude any dead-ends as the workflow process proceeds.

The remaining fields in the state form provide the Processing Options for this state. Processing options are notification choices to signal the various participants about the current status of a document. These e-mail-based signals include customized text for each option.

Table 5.1 lists the field names and types for the *wf_State* form fields, and Figure 5.6 displays the *wf_State* form as seen from a Web browser.

Table 5.1 **The *wf_State* form design.**

Field	Field Type	Field Name
Form type	Text — Computed	wf_FormType
List of forms in the database	Text — Editable	wfFormList
Workflow name	Text — Editable	wf_WorkFlow_Name
Form name	Dialog list of the forms in the database — Editable	wfFormName
State name	Text — Editable	wfs_StateName
State number	Text — Editable	wfs_StateNum
State owner	Dialog list of all role definitions — Editable	wfs_StateOwner
State duration number	Number — Editable	wfs_StateDuration_Num
State duration units	Dialog list — choices: days, weeks, months	wfs_StateDuration_Unit
Send overdue notification?	Radio button — choices: Yes, No	wfs_StateDurExceed
State overdue notice text	Rich Text — Editable	wfs_StateOverdueText
Send overdue follow-up?	Radio button — choices: Yes, No	wfs_SendOverDueFollowup
Overdue follow-up duration	Number — Editable	wfs_FollowDuration_Num
	Dialog List — choices: days, weeks, months	wfs_FollowDuration_Units
Overdue follow-up text	Rich Text — Editable	wfs_OverdueFollowupText

Figure 5.6 The Web browser view of a new *wf_State* form.

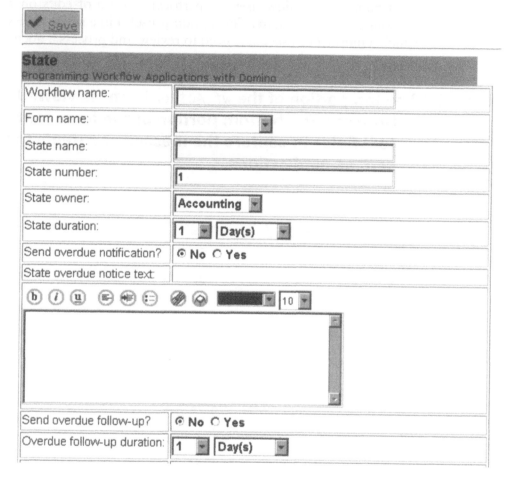

5.3.2 The Action Form

Actions are defined in the *wf_Action* form (Figure 5.7, see page 128) by first assigning a name and number to the action. A set of check boxes allows specification of the states that are associated with this action. That is, if the action can be applied when a document is in a particular state, that state is checked.

Next, we indicate whether this action will cause an automatic state change or if a workflow logic rule (see 5.7 "Constructing the Rule Engine" on page 175) will be evaluated to determine the next state. This is indicated by use of the "State change" or "Apply workflow logic rule" radio buttons.

If we choose the "State change" option, we also have the ability to select "- No Change -" to allow users to perform a series of edits on a document without changing its state. This option is useful in a serial review process in which a number of users may need to review and process a document before the next state can be entered.

Figure 5.7 The top portion of the *wf_Action* form, as viewed in a Web browser. The bottom portion of the form contains rich text fields to capture the message contents that are sent as notifications.

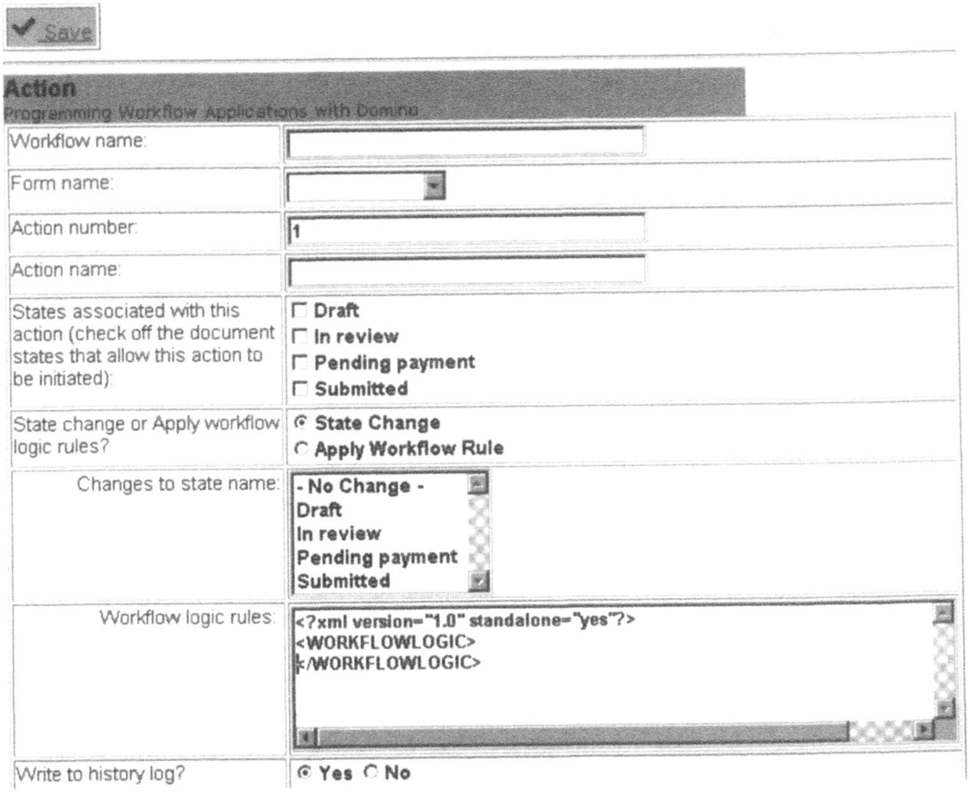

The "Write to history log?" radio button option allows documentation for each action applied to a document. Again, use of this option may be limited to milestone type actions that are needed to maintain an audit trail.

The "Notify state owner" radio button provides for communication throughout the workflow process by optionally sending a customized message to the owner. The message is sent if the document transitions to a new state as a result of the execution of this action.

The "Notify first role member only?" option allows for the primary member of a role that has been identified in a Group document to be the recipient of the state owner notification. This option might be used in a situation in which the state owner is notified of a recent change, primarily as an FYI (For Your Information) message, rather than as a request to perform a specific task.

The "Notify others via e-mail?" field provides additional communication functionality in a workflow process. It optionally causes a customized message to be sent to an individual or group upon the execution of a specific action.

Our last communications-related field is the "Notify originator?" radio button and message. This keeps the document originator in the loop throughout all steps (or actions) in the workflow process.

Table 5.2 The *wf_Action* form design.

Field Name	Field Type	Field Name
Form type	Text — Computed	wf_FormType
List of forms in the database	Text — Editable	wfFormList
Workflow name	Text — Editable	wf_WorkFlow_Name
Action number	Text — Editable	wfa_ActionNumber
Action name	Text — Editable	wfa_ActionName
States associated with action	Dialog list — choices: list of wf_State documents	wfa_ListOfStates
State change or Apply workflow logic rule	Radio button — choices: State Change, Apply Workflow Logic Rule	wfa_StateChangeOrRule
Changes to state name	Dialog list — choices: list of all States	wfa_NewStateName
Workflow logic rule	Rich Text — Editable	wfa_Rule
Write to history log?	Radio button — choices: Yes, No	wfa_WriteHistory
Notify state owner?	Radio button — choices: Yes, No	wfa_NotifyOwner

Table 5.2 The `wf_Action` **form design. (continued)**

Field Name	Field Type	Field Name
Notify first owner role member only?	Radio button — choices: Yes, No	`wfa_NotifyOneOwner`
State owner notification text	Rich Text — Editable	`wfa_OwnerText`
Notify others via e-mail?	Radio button — choices: Yes, No	`wfa_NotifyOthers`
Notify others group or list	Text — Editable	`wfa_NotifyList`
Group to notify	Text — Editable	`wfa_NotifyOthersGroup`
Notify others text	Rich Text — Editable	`wfa_OtherText`
Notify originator	Radio button — choices: Yes, No	`wfa_NotifyOriginator`
Notify originator text	Rich Text — Editable	`wfa_OriginatorText`

5.3.3 The Role Form

`wf_Role` documents describe the actors in a workflow process. In this form, a role is assigned to a specific workflow by completing the "Workflow name" field. Each role document also designates a "Form name", along with a role name and number assigned to the role (Figure 5.8). The check boxes specify all of the valid actions that are available to an actor assigned to the role.

To make the workflow administration flexible, we also specify where the members of this role will be defined. The choices are to define them as a group in the Domino directory, or to define them within the role document itself (thus avoiding changes to the Domino directory). The latter option is helpful in situations in which the group serves no other function outside of the workflow, or if it is problematic within your organization to create and modify Domino groups at the server level where the application resides.

The "Role NAB group" field is used if you select the "Domino Directory" option for the "Role members defined" field. The "Role NAB group" field presents a drop-down list of all the groups existing in your Domino Directory. The "Role members" field is used if you select the "Role Document" option. The user names of the group members are listed in this field, or wild-card characters (e.g., *.*) can be used to designate access to all users for this role. An example of when *.* in the "Role members" field is useful would be a role of "Originator" or "Requester", in which the individuals who could fill this role are not definable at the time the application is configured.

Table 5.3 The *wf_Role* **form design.**

Field	Field Type	Field Name
Form type	Text — Computed	wf_FormType
List of forms in the database	Text — Editable	wfFormList
Workflow name	Text — Editable	wf_WorkFlow_Name
Role name	Text — Editable	wfr_RoleName
Role number	Text — Editable	wfr_RoleNumber
Actions associated with the role	Dialog list — choices: actions from wf_Action documents	wfr_ListOfActions
Role members defined	Radio button — choices: Domino Directory, Role Document	wfr_Defined
Role NAB Group	Text — Editable	wfr_RoleNABGroup
Role members	Text — Editable	wfr_RoleMembers

Figure 5.8 A Web browser view of the *wf_Role* **form.**

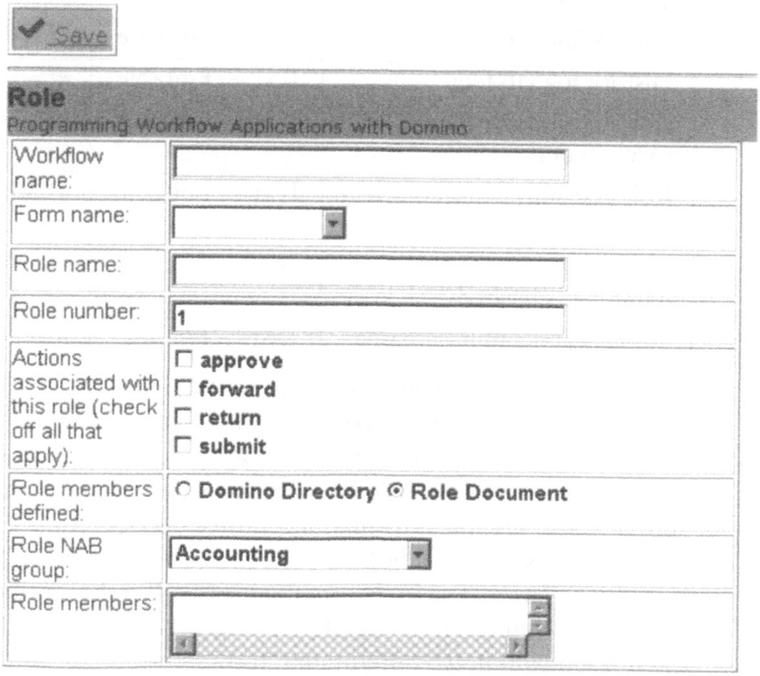

5.3.4 Additional Workflow Design Components

In addition to the three principal state/action/role forms, there are a few ancillary forms that we need to define a basic workflow process. The *wf_ Profile* form provides the definition of a few basic parameters for a workflow process.

First is the workflow administrator access group, used to standardize the administrator access control for forms, views, and fields in an application. Second are two fields on the profile form (Figure 5.9) that specify the server location and database file name for a history log, if one is to be used in the application. Third are two fields to indicate where the Domino directory is located, and the file name of that database. Fourth are the last two fields to specify the server and file name for an instance of the notification database developed in Chapter 4.

Table 5.4 The *wf_Profile* form design.

Field	Field Type	Field Name
Form type	Text — Computed	wf_FormType
Workflow administrator access group	Text — Editable	wfp_AdminGroup
History log server	Text — Editable	wfp_HistoryLogServer
History log DB file:	Text — Editable	wfp_HistoryLogDB
Roles/Groups Domino directory server	Text — Editable	wfp_DomDirServer
Roles/Groups Domino directory database	Text — Editable	wfp_DomDirDatabase
Server where the notification database is located	Text — Editable	wfp_NotificationDBServer
Notification database name	Text — Editable	wfp_NotificationDB

The last part of the design, shared by all five of the forms that define a workflow process, is a subform named *wf_Config*. This subform, with the design summarized in Table 5.5 and the subform illustrated in Figure 5.10 (see page 134), is included at the top of each form design. Its purpose is to supply some session information, mostly derived from CGI variables. The subform consists of four fields: *Server_Name*, *CurrentDB*, *CurUser*, and *disp_ formtype*. The *disp_formtype* field is displayed in a title bar at the top of each document, with the title specified by the *wfFormType* field value included at the top of each workflow configuration form.

Figure 5.9 The workflow process *wf_Profile* **form.**

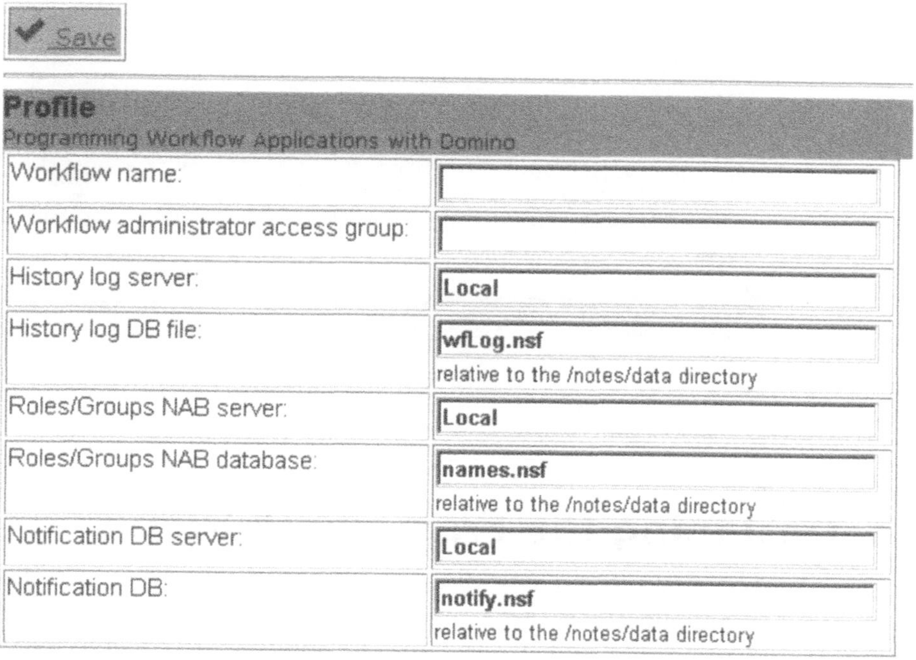

The *wf_Config* subform also includes three action buttons: "Save", "Edit", and "Delete". These are used to manage the workflow configuration documents from the Web.

Table 5.5 The *wf_Config* **form design.**

Field	Computed Value
Server_Name†	Server_Name
CurrentDB	@ReplaceSubstring(@ReplaceSubstring(@Subset(@DbName;-1); "\\";"/"); " "; "+")
CurUser	@Name([CN]; @UserName)
disp_formtype	wf_FormType
† CGI (Common Gateway Interface) Variable	

Figure 5.10 The *wf_Config* **subform as it appears in Notes Designer.**

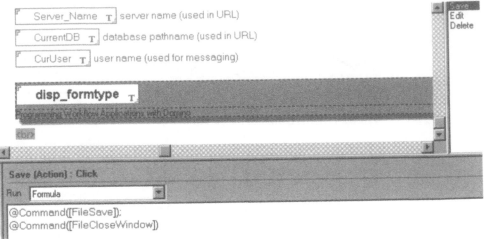

5.4 Designing the Framework Views

The views in our workflow management system, as with most Notes/Domino applications, serve multiple purposes. They let us view selected data that is organized and categorized. In addition, the views will be used by our application code — that is, the LotusScript and Java agents that comprise the engine for our workflow management system.

5.4.1 Viewing Workflow Definitions

The "Workflow\Configurations" view provides us with an overall snapshot of a workflow process. Initially, the view is categorized first on the user-supplied name of the workflow process, and second on the type of workflow document (i.e., action, role, state, or profile). Finally, the specific profile, role, state, and action documents are displayed. Figure 5.11 displays the view in its expanded form. Table 5.6 shows the formulas for the view selection and for each view column.

The action, state, and role views are all similar in design. The "Workflow\Actions" view is categorized on the form type, displaying the action names and the date of the last document modification. The "Workflow\States" and "Workflow\Roles" views are also categorized by form type, displaying the state or role names and last modified dates.

To facilitate the workflow engine processing, we need two additional views for each of the three main configuration document types. These six views are *wfRolesLookupByName*, *wfRolesLookupByNumber*, *wfActionsLookupByName*, *wfActionsLookupByNumber*, *wfStatesLookupByName*, and *wfStatesLookup-ByNumber*. These views allow access to the configuration documents through either the configuration name or its numeric designation.

Table 5.6 The "Workflow\Configurations" view design.

Name	Title	Formula
View	Workflow\Configurations	SELECT (Form="wf_Profile") | (Form="wf_State") | (Form="wf_Action") | (Form="wf_Role")
Column 1	Workflow process	wf_WorkFlow_Name
Column 2	Form type	wf_FormType
Column 3	Action/Role/State name	@If(Form = "wf_State"; wfs_StateName; Form = "wf_Action"; wfa_ActionName;Form = "wf_Role"; wfr_RoleName; Form = "wf_Profile"; "Profile";"Error")

Figure 5.11 A sample "Workflow\Configurations" view.

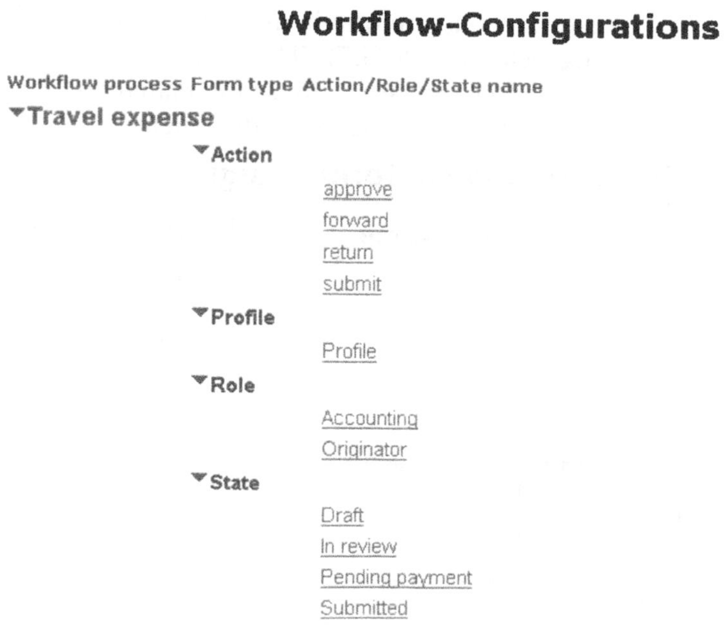

Workflow-Configurations

Workflow process Form type Action/Role/State name

▼Travel expense

   ▼Action

      approve
      forward
      return
      submit

   ▼Profile

      Profile

   ▼Role

      Accounting
      Originator

   ▼State

      Draft
      In review
      Pending payment
      Submitted

5.4.2 Designing the Workflow-Enabling Subform

To workflow-enable a form in a database, we need to calculate, capture, and collate the workflow information for that form's documents as they proceed along their lifecycles. To create a portable workflow solution that can be added to a form, we create a workflow subform. This subform can handle all workflow-related processing without disturbing or relying on features of the main form.

Our subform identifies what state the document is in, and what role(s) the current user assumes in the workflow. Based on those pieces of information, the subform can calculate what actions are available. However, we want the subform to provide additional information, such as how long a document may reside in this state, who is the current owner of the document, and if any late notices have already been sent. Also, we want to create an audit trail for each workflow-enabled document so we can quickly review the progress of a specific document without resorting to a system log file. For this functionality, we rely on the *wfHistory* field. Finally, we need to dynamically present to the user only those actions that are valid for the combination of their particular role(s) and the document's current state. This is accomplished by running an agent in the *wfCurValidActions* field when the enclosing form is loaded.

Let's review this subform in more detail. The subform is comprised of the following fields (Table 5.7). The first three fields (current server, current database, and current user) are self-explanatory. These fields make this session information available to the document.

Table 5.7 The workflow subform design.

Field	Field Name	Field Type	Formula
Current server	wfCurrentServer	Text — CFD	@Subset(@DbName;1)
Current database	wfCurrentDB	Text — CFD	@ReplaceSubstring(@ReplaceSubstring(@Subset(@DbName;-1); "\\";"/"); " "; "+")
Current user	wfCurrentUser	Text — CFD	@Name([CN]; @UserName)
Form name for the document	wfFormName	Text — CWC	wf_FormType handles the need for a form name before a document is initially saved
Roles of the current user	wfRole	Text List — Editable	computed by WebQueryOpen agent wfGetCurActionsList

Table 5.7 The workflow subform design. (continued)

Field	Field Name	Field Type	Formula
Previous state number	wfPrevState_num	Number — Editable	computed by WebQuerySave agent wfProcess
Current user name (saved)	wfSavedUser	Text — Editable	Default Value = wfCurrentUser
Originator of the document	wfOriginator	Text — CWC	wfCurrentUser
State number	wfState_num	Number — Editable	Default Value = 1
State name	wfState_name	Text — Computed	@DbLookup("":"NoCache";""; "Workflow\\ wfStatesLookupByNumber"; wfState_num;2)
State owner	wfOwner	Text — CFD	@DbLookup("":"NoCache";""; "Workflow\\ wfStatesLookupByNumber"; wfState_num;3)
Current list of valid actions	wfCurValidActions	Text List — Editable	computed by WebQueryOpen agent wfGetCurActionsList
Last action date	wfLastActionDate	Date/Time — Computed	@If(@IsNewDoc;@Today; wfLastActionDate)
Next action date	wfNextActionDate	Date/Time — Editable	wfNextActionDate — computed by WebQuerySave agent wfProcessForm
Last notice sent date	wfLateNoticeSentDate	Date/Time — Editable	
History of the workflow actions	wfHistory	Text List — Editable	Computed by WebQuerySave agent wfProcessForm
Actions the current user may carry out	wfActions	Radio Button — Editable	@Trim(wfCurValidActions)
CFD — Computed For Display CWC — Computed When Composed			

The *wfFormName* field computes the value of the current document's form, based on a field named *wfFormType* that is defined at the top of the form before the subform is added.

The *wfRole* field holds the list of valid roles for the current user. The *wfCurValidActions* field is another field that contains a list of the valid actions for a user, based on his/her roles and the document's current state. Both of these fields are set by the *WebQueryOpen* agent named *wfGetCurActionsList*. This agent is described in 5.5 "Creating the Web Agents" on page 139.

Both the previous state and the current user are saved in the next two fields, *wfPrevState_num* and *wfSavedUser*. The current state number defaults to 1 for a newly created document. It is updated thereafter by the workflow engine.

The originator and owner fields are computed. The originator is computed when composed to equal the author of the document. The owner field is computed by taking the current state number of the document and looking up the particular state owner from the appropriate *wf_State* form. Note that the state name and owner use the *wfStatesLookupByNumber* view to find the appropriate state document.

The last action date defaults to the current date for a newly created document. It is updated throughout the lifecycle of the workflow process whenever processing takes place. The next action date is computed by the *wfProcessForm* agent that is called when the document is saved. The agent calls a script library function to look up the duration period for a state when a document enters that state. The agent then calculates the due date for the next action.

The late notice sent date is utilized if a late notice has been sent for a document. Follow-up notices can be sent according to the schedule specified in the state document.

The *wfHistory* field is used to provide an audit trail on the workflow processing of each particular document.

Finally, the *wfActions* field is a computed radio button that provides the current user with only the actions that their particular role can perform on a document in its current state. The actions are obtained from the value of the *wfCurValidActions* field that is populated when the document loads.

5.5 Creating the Web Agents

Our workflow management system uses several agents (Table 5.8) and a script library for basic processing of a workflow-enabled document.

Table 5.8 Agents used in the workflow management system.

Agent Name	Associated Form	Type of Agent	Language
wfGetCurActionsList	wf_Subform	WebQueryOpen	LotusScript
wfGetFormList	wf_Action wf_State wf_Role	WebQueryOpen	LotusScript
wfProcess	wf_Subform	WebQuerySave	Java
wfProcessForm	wf_Subform	WebQuerySave	LotusScript

5.5.1 The *wfGetCurActionsList* Agent

This agent (Listing 5.1) is used in the *WebQueryOpen* event of the "Sample Form" form in the example database. It searches all of the defined roles for this workflow to see which ones are assigned to the current user. This list is used to set the value of the *wfRole* field in the subform. The agent then looks up the actions available to each role, and forms a complete list of actions that a user with those roles can perform. Finally, this list of actions is filtered by the current document's state. That is, if the action cannot be performed while the document is in this state, the action is not included in the list. Thus, the final list of actions consists of all valid actions for this user and this state. These actions are set into the *wfCurValidActions* field.

Listing 5.1 The LotusScript code for the *wfGetCurActionsList* agent.

```
'****************************************************************************
'
' Module: wfGetCurActionsList.ls
' Author: Dick Lam
'
' Description: This agent computes the allowable actions that can take place,
'              based on the current state of a document in the current
'              workflow process. It is run as a WebQueryOpen agent in the
'              SampleForm form. The agent uses the WorkflowDocument class
```

Listing 5.1 The LotusScript code for the *wfGetCurActionsList* **agent. (continued)**

```
'                    defined in the wfScriptLibrary.
'
'****************************************************************************


Sub Initialize
    Dim s As NotesSession, db As NotesDatabase, note As NotesDocument
    Dim wf As WorkflowDocument, sUser As String
    Dim roleView As NotesView, actionView As NotesView
    Dim roleDoc As NotesDocument, actionDoc As NotesDocument
    Dim docActions As NotesItem, docStates As NotesItem
    Dim roles As Variant, curState As Variant, actionList() As String
    Dim numActions As Integer, actions As Variant, states As Variant
    Dim addAction As Variant, index As Variant
    On Error Goto wfGetCurActionsListError

    ' get the current session parameters
    Set s = New NotesSession
    Set db = s.CurrentDatabase
    Set note = s.DocumentContext

    ' access the role and action views
    Set roleView = db.GetView("Workflow\wfRolesLookupByName")
    Set actionView = db.GetView("Workflow\wfActionsLookupByName")

    ' create a WorkflowDocument instance
    Set wf = New WorkflowDocument(s, db, note)

    ' get the current user and the roles he/she belongs to
    sUser = wf.getThisUser
    roles = wf.getRoles(sUser)
    Call note.ReplaceItemValue( "wfRole", Fulltrim(roles) )
```

Listing 5.1 The LotusScript code for the *wfGetCurActionsList* agent. (continued)

```
' get the current document state name
curState = wf.getStateName


' for each user role, get the actions that are associated with it - filter
' the actions based on the current document state, and store the list of
' available actions in the actionList array
numActions = 0
Redim actionList(numActions)


Forall r In roles
    ' r is the current role name - lookup the role document
    If (r = "") Then
        Set roleDoc = Nothing
    Else
        Set roleDoc = roleView.GetDocumentByKey(r)
    End If
    If Not (roleDoc Is Nothing) Then
        ' get all actions that are associated with the role
        Set docActions = roleDoc.GetFirstItem("wfr_ListOfActions")
        actions = docActions.Values


        ' are there any actions to process?
        If Not Isnull(actions) Then
            ' iterate through all actions associated with this role
            Forall a In actions
                ' lookup the corresponding action document
                Set actionDoc = actionView.GetDocumentByKey(a)


                ' got the action document?
                If Not (actionDoc Is Nothing) Then
                    ' see if the action is associated with the current state
                    Set docStates = actionDoc.GetFirstItem("wfa_ListOfStates")
```

Listing 5.1 The LotusScript code for the *wfGetCurActionsList* agent. (continued)

```
            states = docStates.Values

            ' any states associated with this action?
            If Not Isnull(states) Then
                index = Arraygetindex(states, curState)
                If Not Isnull(index) Then
                    ' action is valid for this state - if
                    ' the action is not a duplicate, add
                    ' it to the action list array
                    addAction = True
                    Forall v In actionList
                        If (v = a) Then
                            addAction = False
                        End If
                    End Forall
                    If addAction Then
                        ' redimension the array
                        If (numActions = 0) Then
                            Redim actionList(1)
                        Else
                            Redim Preserve actionList(numActions + 1)
                        End If

                        ' add the new action to the list
                        actionList(numActions) = a
                        numActions = numActions + 1
                    End If
                End If
            End If
        End Forall
End If
```

Listing 5.1 The LotusScript code for the *wfGetCurActionsList* **agent. (continued)**

```
        End If
    End Forall


    ' now that we have the actions available to this user for a document
    ' in this state, write these actions to the workflow subform
    If (numActions = 0) Then Goto wfGetCurActionsListError
    Call note.ReplaceItemValue("wfCurValidActions", actionList)
    Call note.ReplaceItemValue("wfActions", actionList(0))
    Exit Sub


wfGetCurActionsListError:
    Call note.ReplaceItemValue("wfCurValidActions", "None")' default action
    Call note.ReplaceItemValue("wfActions", actionList(0))
    Exit Sub
End Sub
```

5.5.2 The *wfProcess* Agent

The *wfProcess* agent runs in the *WebQuerySave* event of a workflow-enabled document. This is a Java agent that is fired when a workflow document is saved. It reads the action document and either changes the state of the document, or launches the rule engine (covered in 5.7 "Constructing the Rule Engine" on page 175) to decide the next state. It then chains to the *wfProcessForm* agent to carry out other tasks associated with the action (e.g., notifications).

Note that the chaining of the second agent from *wfProcess* is accomplished using a new feature of the Notes API, beginning with Notes/Domino version 5.0.2. This feature allows one agent to launch another one, passing it the note ID for a document. The agent that is launched receives the note ID and uses it to retrieve the document and access any shared variables between the two agents. Also, it is worth noting that there are some problems in 5.0.2 that caused us to upgrade to version 5.0.2a as soon as that maintenance release became available.

Listing 5.2 The *actionTask* **class of the** *wfProcess* **agent.**

```java
// ******************************************************************
//
// Module: actionTask.java
// Author: Dick Lam
//
// Description: This is a Java class that implements the notesTask interface.
//              The task of this class is to determine the next state
//              change for the document and then chain to the wfProcessForm
//              agent.
//
// ******************************************************************

import java.io.*;
import java.net.*;
import java.util.*;
import lotus.domino.*;

// ******************************************************************

public class actionTask implements notesTask
{
    // constructor
    public actionTask()
    {
    }

    // -----------------------------------------------------------------

    // perform - performs this task
    public void perform()
    {
```

Listing 5.2 The *actionTask* **class of the** *wfProcess* **agent. (continued)**

```
try {
    // retrieve the document from the agent context and get the
    // form name, selected action (so the appropriate action document
    // can be retrieved), and the current state
    Document doc = wfProcess.getContext().getDocumentContext();
    String formName = doc.getItemValueString("wfFormName");
    String actionName = doc.getItemValueString("wfActions");
    int originalState = doc.getItemValueInteger("wfState_num");
    String user = doc.getItemValueString("wfCurrentUser");

    // save the original state, user, and role
    Item i = doc.getFirstItem("wfPrevState_num");
    i.setValueInteger(originalState);
    doc.replaceItemValue("wfSavedUser", user);

    // get the action document from the actions view
    View actionView =
      wfProcess.getDatabase().getView("Workflow\\wfActionsLookupByName");
    Document actionDoc =
      actionView.getDocumentByKey(actionName);

    // determine if this action is a direct state change or if a
    // rule or rules must be applied to change the state
    String actionType =
      actionDoc.getItemValueString("wfa_StateChangeorRule");
    if ( actionType.equalsIgnoreCase("State") ) {
      // direct state change - get the new state name and number
      String newStateName =
        actionDoc.getItemValueString("wfa_NewStateName");
      View statesView =
        wfProcess.getDatabase().getView("Workflow\\wfStatesLookup-
ByName");
      Document stateDoc =
```

Listing 5.2 The *actionTask* **class of the** *wfProcess* **agent. (continued)**

```
            statesView.getDocumentByKey(newStateName);
        int newStateNum = stateDoc.getItemValueInteger("wfs_StateNum");


        // update the state number in the document being processed
        Item j = doc.getFirstItem("wfState_num");
        j.setValueInteger(newStateNum);
    } else {
        // launch the rule engine
        String rules = actionDoc.getItemValueString("wfa_Rule");
        wfLogic wf = new wfLogic();
        wf.load( new StringReader(rules) );


        // check for errors
        if ( wf.anyErrors() )
            notesLogFile.getInstance().write("errors detected!");
        else
            notesLogFile.getInstance().write("no errors or warnings
detected!");
        if ( wf.isUsable() ) {
            wf.run(doc);
        }
        wf = null;
    }


    // save the document and chain to the next agent (wfProcessForm)
    // (note: you must be using Domino 5.0.2 or greater for this
    // agent chaining code to work)
    doc.computeWithForm(false, false);
    doc.save();
    Agent nextAgent = wfProcess.getDatabase().getAgent("wfProcessForm");
    nextAgent.run( doc.getNoteID() );
    }
    catch (Exception e) {
```

Listing 5.2 The *actionTask* **class of the** *wfProcess* **agent. (continued)**

```
                notesLogFile.getInstance().write( "actionTask#perform(): " +
                                          e.toString() );
        }
    }
}
// **************************************************************************
// end of actionTask.java
```

5.5.3 The *wfProcessForm* Agent

The *wfProcessForm* agent (Listing 5.3) is launched by the *wfProcess* agent to complete the processing on the document. This agent updates the history field and initiates any notifications through the notification engine.

Listing 5.3 The LotusScript code for the *wfProcessForm* **agent.**

```
'**************************************************************************
'
' Module: wfProcessForm.ls
' Author: Dan Giblin
'
' Description: This agent is launched by wfProcess to finish the WebQuerySave
'              processing of a workflow-enabled document. The agent uses the
'              WorkflowDocument class defined in the wfScriptLibrary.
'
'**************************************************************************

Sub Initialize
    Dim s As NotesSession, db As NotesDatabase, note As NotesDocument
    Dim agent As NotesAgent, paramid As String
    Dim wf As WorkflowDocument
    On Error Goto wfProcessFormError

    ' get the session parameters
    Set s = New NotesSession
```

Listing 5.3 The LotusScript code for the *wfProcessForm* agent. (continued)

```
Set db = s.CurrentDatabase
Set agent = s.CurrentAgent

' get the note ID passed in by the agent that launched this one
' (note: this is a Domino v5.0.2 feature)
paramid = agent.ParameterDocID

' retrieve the note based on its ID, and use it to create a
' new instance of a WorkflowDocument
Set note = db.GetDocumentByID(paramid)
Set wf = New WorkflowDocument(s, db, note)

' process the document and return
Call wf.process
Exit Sub

wfProcessFormError:
    Exit Sub
End Sub
```

wfScriptLibrary is a LotusScript script library used by the *wfGetCurActionsList* and *wfProcessForm* agents. It contains three LotusScript classes (*WorkflowDocument*, *Notification*, and *notesLogFile*) containing properties and methods for the manipulation of workflow-enabled documents.

The *WorkflowDocument* class (Listing 5.4) was designed to provide access to some of the fields in the workflow subform, and provide a *process()* method for handling the post-processing after the state change has been completed by *wfProcess*. It also includes a *getRoles()* method that enumerates all roles that are valid for a particular user.

Listing 5.4 The *WorkflowDocument* class of *wfScriptLibrary*.

```
'*****************************************************************************
'
' Module: WorkflowDocument.ls
' Author: Dan Giblin and Dick Lam
'
' Description: This LotusScript class handles processing of actions on
'              workflow-enabled documents. This class assumes that the
'              document has been pre-processed by the wfProcess agent.
'
'*****************************************************************************


Public Class WorkflowDocument
    ' properties

    Private curSession As NotesSession    ' initialized in the constructor
    Private curDB As NotesDatabase
    Private curDoc As NotesDocument


    ' values extracted from the current document
    Private sAction As String        ' current action
    Private sActionNum As String
    Private sOriginator As String        ' document originator
    Private sCurUser As String        ' current user
    Private sOrigState As String        ' previous state
    Private sNewStateName As String    ' current state
    Private iNewStateNum As Integer
    Private sNewOwner As String        ' owner of current state
    Private sForm As String            ' form name of document


'*****************************************************************************
```

Listing 5.4 The *WorkflowDocument* **class of**
           wfScriptLibrary. **(continued)**

```
' constructor
Sub New(s As NotesSession, db As NotesDatabase, doc As NotesDocument)
    ' save the arguments in this instance
    Set curSession = s
    Set curDB = db
    Set curDoc = doc


        ' extract the document values
    Call getDocValues
End Sub


'----------------------------------------------------------------------


' process - carries out the standard workflow tasks on the current
'           document (returns True if ok, False if not)
Function process As Variant
    Dim wfDate As NotesDateTime, actionDoc As NotesDocument
    Dim actionView As NotesView, writeTheHistory As Variant
    Dim i As NotesItem
    On Error Goto processError


    ' see what action was requested
    If sAction = "None" Then
        ' "None" is the action if a user tries to process a form that they
        ' can still see, but there are no valid workflow options for them
        process = False
        Exit Function
    Else
        ' retrieve the action document and see if the history should
        ' be written to the current document
        writeTheHistory = False
```

Listing 5.4 The *WorkflowDocument* **class of**
wfScriptLibrary. **(continued)**

```
            On Error Goto historyFlagLookupError
            Set actionView = curDB.GetView("Workflow\wfActionsLookupByName")
            Set actionDoc = actionView.GetDocumentByKey(sAction)
            Set i = actionDoc.GetFirstItem("wfa_WriteHistory")
            If (i.Values(0) = "1") Then
                writeTheHistory = True
            End If
historyFlagLookupError:
            On Error Goto processError
        End If


        ' set the last action date to today
        Set wfDate = New NotesDateTime("Today")
        Set curDoc.wfLastActionDate = wfDate


        ' set the next action date to today plus the state duration
        Me.setNextActionDate(iNewStateNum)


        ' write the action to the history log and save the document
        If writeTheHistory Then
            Call Me.setHistory
            Call Me.wfSave
        End If


        ' carry out the processing options and exit
        Call processOptions
        process = True
        Exit Function

processError:
        process = False
    End Function
```

Listing 5.4 The *WorkflowDocument* **class of**
**   *wfScriptLibrary*. (continued)**

```
    '-------------------------------------------------------------------

    ' processOptions - carries out notifications based on the notification
    '                  parameters in the current document
    Sub processOptions
        Dim notifier As Notification, key As String
        On Error Goto processOptionsError


        ' create a notification object
        key = sAction
        Set notifier = New Notification(curDB, key, sNewOwner, sOriginator)
        Call notifier.notifyAll
        Exit Sub


processOptionsError:
        Exit Sub
    End Sub


    '-------------------------------------------------------------------


    ' setNextActionDate - computes the interval before the next action
    '                     based on the duration stored in the state document
    Sub setNextActionDate(state As Integer)
        Dim num As Integer, units As String, dt As NotesDateTime
        Dim v As NotesView, doc As NotesDocument, i As NotesItem
        On Error Goto setNextActionDateError


        ' lookup the state document by number
        Set v = curDB.GetView("Workflow\wfStatesLookupByNumber")
        Set doc = v.GetDocumentByKey(state)
```

Listing 5.4 The *WorkflowDocument* **class of**
        wfScriptLibrary. **(continued)**

```
        ' get the duration number and duration units
    Set i = doc.GetFirstItem("wfs_StateDuration_Num")
    num = i.Values(0)
    Set i = doc.GetFirstItem("wfs_StateDuration_Unit")
    units = i.Values(0)


      ' compute the next action date by adding the state duration
      ' to the current date
    Set dt = New NotesDateTime("Today")
    If (units = "1") Then
       dt.AdjustDay(num)        ' days
    Elseif (units = "2") Then
       dt.AdjustDay(num*7)      ' weeks
    Elseif (units = "3") Then
       dt.AdjustMonth(num)       ' months
    End If


      ' set the next action date field
    Set curDoc.wfNextActionDate = dt
    Exit Sub


setNextActionDateError:
    Exit Sub
  End Sub


  '-------------------------------------------------------------------------


   ' getHistory - returns the current document's history field contents
  Function getHistory As Variant
    Dim i As NotesItem
    On Error Goto getHistoryError
```

Listing 5.4 The *WorkflowDocument* **class of** *wfScriptLibrary*. **(continued)**

```
      Set i = curDoc.GetFirstItem("wfHistory")
      getHistory = i.Values
      Exit Function


getHistoryError:
      getHistory = ""
    End Function



      '-------------------------------------------------------------------------



      ' setHistory - logs an action to the history field
      Sub setHistory
        Dim logentry As String, dt As NotesDateTime, vNow As Variant
        Dim i As NotesItem
        On Error Goto setHistoryError


          ' get the current time
        Set dt = New NotesDateTime("Today")
        Call dt.SetNow
        vNow = dt.LSLocalTime


          ' create the string summarizing the action
        logentry = "Time = " & vNow & _
        ": Action = " & sAction & _
        " | Original state = " & sOrigState & _
        " | New state = " & sNewStateName


          ' append the new string to the current history field contents
        Set i = curDoc.GetFirstItem("wfHistory")
        i.AppendToTextList(logentry)
        Exit Sub
```

Listing 5.4 The *WorkflowDocument* **class of**
          wfScriptLibrary. **(continued)**

```
setHistoryError:
    Exit Sub
End Sub


'-------------------------------------------------------------------


' wfSave - saves the current document
Sub wfSave
    On Error Resume Next
    Call curDoc.Save(True, True)
End Sub


'-------------------------------------------------------------------


' getDocValues - extracts the document parameters
Private Sub getDocValues
    On Error Resume Next

    ' get the current document parameters
    sAction = Me.getSelectedAction
    sActionNum = Me.getSelectedActionNum
    sOriginator = Me.getOriginator
    sCurUser = Me.getThisUser
    sOrigState = Me.getPrevStateName
    sNewStateName = Me.getStateName
    iNewStateNum = Me.getStateNum
    sNewOwner = Me.getNewOwner
    sForm = Me.getFormName
    Exit Sub

getDocValuesError:
    Exit Sub
End Sub
```

Listing 5.4 The *WorkflowDocument* **class of**
       wfScriptLibrary. **(continued)**

```
'------------------------------------------------------------------


  ' getSelectedAction - returns the action name
  Function getSelectedAction As String
     On Error Goto getSelectedActionError
     getSelectedAction = curDoc.wfActions(0)
     Exit Function


getSelectedActionError:
     getSelectedAction = "None"
  End Function



    '----------------------------------------------------------------


   ' getSelectedActionNum - returns the action number
  Function getSelectedActionNum As Integer
     Dim v As NotesView, doc As NotesDocument, i As NotesItem
     On Error Goto getSelectedActionNumError


     Set v = curDB.GetView("Workflow\wfActionsLookupByName")
     Set doc = v.GetDocumentByKey(Me.getSelectedAction)
     Set i = doc.GetFirstItem("wfa_ActionNumber")
     getSelectedActionNum = i.Values(0)
     Exit Function


getSelectedActionNumError:
     getSelectedActionNum = 0
  End Function
```

Listing 5.4 The *WorkflowDocument* **class of**
          wfScriptLibrary. **(continued)**

```
    '----------------------------------------------------------------------

        ' getOriginator - returns the document originator
        Function getOriginator As String
           On Error Goto getOriginatorError
           getOriginator = curDoc.wfOriginator(0)
           Exit Function


    getOriginatorError:
           getOriginator = ""
       End Function



        '--------------------------------------------------------------------



        ' getThisUser - returns the current user and saves the user name in the
        '               wfSavedUser field
        Function getThisUser As String
           Dim user As String
           On Error Goto getThisUserError

           user = curDoc.wfCurrentUser(0)
           curDoc.wfSavedUser = user
           getThisUser = user
           Exit Function


    getThisUserError:
           getThisUser = ""
       End Function
```

Listing 5.4 The *WorkflowDocument* **class of**
          wfScriptLibrary. **(continued)**

```
'------------------------------------------------------------------

    ' getPrevStateName - returns the previous state of the document
    Function getPrevStateName As String
        Dim v As NotesView, doc As NotesDocument, i As NotesItem
        On Error Goto getPrevStateNameError


        Set v = curDB.GetView("Workflow\wfStatesLookupByNumber")
        Set doc = v.GetDocumentByKey( curDoc.wfPrevState_num(0) )
        Set i = doc.GetFirstItem("wfs_StateName")
        getPrevStateName = i.Values(0)
        Exit Function


getPrevStateNameError:
        getPrevStateName = ""
    End Function


    '------------------------------------------------------------------


    ' getStateName - returns the current state of the document
    Function getStateName As String
        On Error Goto getStateNameError
        getStateName = curDoc.wfState_Name(0)
        Exit Function


getStateNameError:
        getStateName = ""
    End Function
```

Listing 5.4 The *WorkflowDocument* **class of**
wfScriptLibrary. **(continued)**

```
    '-------------------------------------------------------------------


    ' getStateNum - returns the current state number of the document
    Function getStateNum As Integer
        On Error Goto getStateNumError
        getStateNum = curDoc.wfState_num(0)
        Exit Function


getStateNumError:
        getStateNum = 0
    End Function



    '-------------------------------------------------------------------


    ' getNewOwner - returns the owner for the current state
    Function getNewOwner As String
        Dim v As NotesView, doc As NotesDocument, i As NotesItem
        On Error Goto getNewOwnerError


        Set v = curDB.GetView("Workflow\wfStatesLookupByNumber")
        Set doc = v.GetDocumentByKey(Me.getStateNum)
        Set i = doc.GetFirstItem("wfs_StateOwner")
        getNewOwner = i.Values(0)
        Exit Function


getNewOwnerError:
        getNewOwner = ""
    End Function
```

Listing 5.4 The *WorkflowDocument* **class of**
**   *wfScriptLibrary*. (continued)**

```
'----------------------------------------------------------------


    ' getFormName - returns the form name of the document
    Function getFormName As String
        On Error Goto getFormNameError
        getFormName = curDoc.wfFormName(0)
        Exit Function


getFormNameError:
        getFormName = ""
    End Function



    '----------------------------------------------------------------


    ' getRoles - retrieves the roles associated with a particular user
    Function getRoles(username As String) As Variant
        Dim nab As NotesDatabase, groupView As NotesView
        Dim roleView As NotesView, iMembers As NotesItem
        Dim roleDoc As NotesDocument, iName As NotesItem, iType As NotesItem
        Dim groupDoc As NotesDocument, iList As NotesItem
        Dim sRoleName As String, sGroupName As String
        Dim allRoles() As String
        Dim count As Integer

        ' access the Domino directory for looking up groups
        Set nab = curSession.GetDatabase("", "names.nsf")
        Set groupView = nab.GetView("Groups")
```

Listing 5.4 The *WorkflowDocument* **class of**
            wfScriptLibrary. **(continued)**

```
                ' lookup the roles in each workflow role definition document
                ' (note that allRoles is the returned array of roles found
                ' for the user)
Set roleView = curDB.GetView("Workflow\Roles")
Set roleDoc = roleView.GetFirstDocument
count = 0
Redim Preserve allRoles(count)

Do While Not (roleDoc Is Nothing)
            ' retrieve the role name and type
    Set iName = roleDoc.GetFirstItem("wfr_RoleName")
    sRoleName  = iName.Values(0)
    Set iType = roleDoc.GetFirstItem("wfr_Defined")

    ' check the type
    If (iType.Values(0) = "Document") Then
        ' check the user's name against the members
        ' in the role document
        Set iMembers = roleDoc.GetFirstItem("wfr_RoleMembers")
        If (iMembers.Contains(username) Or _
        iMembers.Contains("*.*") ) Then
                ' add the current role name to the list of roles
                ' that will be returned
            If (count = 0) Then
                Redim allRoles(1)
            Else
                Redim Preserve allRoles(count + 1)
            End If
            allRoles(count) = sRoleName
            count = count + 1
        End If
    Else
```

Listing 5.4 The *WorkflowDocument* **class of**
        wfScriptLibrary. **(continued)**

```
         ' get the NAB group name
         Set iList = roleDoc.GetFirstItem("wfr_RoleNABGroup")
         sGroupName = iList.Values(0)


             ' retrieve the group document
         Set groupDoc = groupView.GetDocumentByKey(sGroupName)
         If Not (groupDoc Is Nothing) Then
              ' check to see if the user is listed in the group
           Set iMembers = groupDoc.GetFirstItem("Members")
           If (iMembers.Contains(username) Or _
           iMembers.Contains("*.*") ) Then
                       ' add the current role name to the list of roles
                       ' that will be returned
              If (count = 0) Then
                 Redim allRoles(1)
              Else
                 Redim Preserve allRoles(count + 1)
              End If
              allRoles(count) = sRoleName
              count = count + 1
            End If
         End If
      End If


      ' process the next document
      Set roleDoc = roleView.GetNextDocument(roleDoc)
   Loop


      ' return the array of roles
   getRoles = allRoles
   Exit Function
```

Listing 5.4 The *WorkflowDocument* **class of** *wfScriptLibrary*. **(continued)**

```
getRolesError:
    getRoles = allRoles
  End Function


  '-------------------------------------------------------------------


  ' setRole - sets the role list of the user in the current document
  Sub setRole(aRole As Variant)
    On Error Resume Next
    curDoc.wfRole = aRole
  End Sub
End Class
```

The *notesLogFile* class (Listing 5.5) is simply a port of the Java class introduced in Chapter 4. It is a LotusScript version that allows creation of a log file and the ability to write messages to the log. Unfortunately, limitations in the LotusScript language preclude a complete port of *notesLogFile* with all features intact. Some features that do not port between the two languages are LotusScript's lack of a private constructor and no support for static methods within a class.

Listing 5.5 The *notesLogFile* **class of** *wfScriptLibrary*.

```
'**************************************************************************
'
' Module: notesLogFile.ls
' Author: Dick Lam
'
' Description: This LotusScript class implements an instance of a NotesLog
'              class for logging agent messages to an output file.
'
' Usage: Once an agent has created a Session instance, create an instance
'        of this class and write to the log through the write() method.
'
```

Listing 5.5 The *notesLogFile* **class of** *wfScriptLibrary*. **(continued)**

```
'        Dim s as NotesSession, log as notesLogFile
'        Set s = New NotesSession
'        Set log = New notesLogFile(s, "log", "agent.log")
'        Call log.write("This is a sample debug message.")
'

'**************************************************************************


Public Class notesLogFile
   ' properties

   ' NotesLog instance and flag for turning debug messaging on/off
   myLog As NotesLog
   debug As Variant


   '**************************************************************************


   ' constructor
   Sub New(s As NotesSession, logName As String, logFile As String)
      On Error Goto NewError

      ' initialize the properties
      debug = True

        ' create a NotesLog instance
      Set myLog = s.CreateLog(logName)
      myLog.OverwriteFile = True
      Call myLog.OpenFileLog(logFile)
      Exit Sub

NewError:
      ' constructor error
   End Sub
```

Listing 5.5 The *notesLogFile* **class of** *wfScriptLibrary*. **(continued)**

```
'---------------------------------------------------------------


    ' destructor
    Sub Delete
        On Error Goto deleteError


        ' close and delete the NotesLog
        Call myLog.Close
        Set myLog = Nothing
        Exit Sub

deleteError:
            ' destructor error
        End Sub


    '---------------------------------------------------------------


    ' setDebugState - turns debugging log on/off
    Sub setDebugState(state As Variant)
        debug = state
    End Sub


    '---------------------------------------------------------------


    ' write - writes a message to the log if it is active
    Sub write(message As String)
        On Error Goto writeError


        If (debug) Then
            ' write to the log file
            Call myLog.LogAction(message)
        End If
        Exit Sub
```

Listing 5.5 The *notesLogFile* **class of** *wfScriptLibrary*. **(continued)**

```
writeError:
        ' I/O error
    End Sub
End Class
```

5.6 Integrating the Notification Engine

The notification database developed in Chapter 4 is used by the workflow engine for generating outgoing messages that result from state transitions and overdue document instances.

During the processing of the document by the *wfProcessForm* agent, the action document is checked to see if any notifications need to be generated to the originator, state owner, or any other actors. If notifications are requested for any of those situations, the appropriate method in the *Notification* class (Listing 5.6) creates a "Notification Entry" document in the notification database, which subsequently processes the entry according to the specified schedule.

Listing 5.6 The *Notification* **class of** *wfScriptLibrary*.

```
'*****************************************************************************
'
' Module: Notification.ls
' Author: Dan Giblin and Dick Lam
'
' Description: This LotusScript class handles notifications by providing
'              an interface to the Workflow Notification database. The
'              notifications are generated from information in an action
'              document that is retrieved from the current database using
'              a key passed to the object's constructor.
'
'
'*****************************************************************************


Public Class Notification
    ' properties
```

Listing 5.6 The *Notification* **class of** *wfScriptLibrary.* **(continued)**

```
curDB As NotesDatabase      ' from the constructor
curKey As String
sOwner As String
sOriginator As String

curDoc As NotesDocument     ' from a lookup of the action document
sForm As String             ' the form name for the workflow document


'*************************************************************************


' constructor
Sub New(db As NotesDatabase, key As String, owner As String, _
originator As String)
   Set curDB = db
   curKey = key
   sOwner = owner
   sOriginator = originator
End Sub


'-------------------------------------------------------------------------


' notifyAll - processes notifications
Sub notifyAll
   On Error Goto notifyAllError


   ' lookup the action document
   Set curDoc = lookupAction
   sForm = curDoc.wfFormName(0)

   ' notify the state owner
   If Me.getNotifyStateOwner = 1 Then
      Call notifyStateOwner
   End If
```

Listing 5.6 The *Notification* **class of** *wfScriptLibrary*. **(continued)**

```
        ' notify the originator
        If Me.getNotifyOriginator = 1 Then
            Call notifyOriginator
        End If


        ' notify others
        If Me.getNotifyOthers = 1 Then
            Call notifyOthers
        End If
        Exit Sub

notifyAllError:
        Exit Sub
    End Sub


    '--------------------------------------------------------------------


    ' getNotifyOthers - retrieves the flag for specifying notification
    '                   of others
    Function getNotifyOthers As String
        Dim i As NotesItem
        On Error Goto getNotifyOthersError

        Set i = curDoc.GetFirstItem("wfa_NotifyOthers")
        getNotifyOthers = i.Values(0)
        Exit Function

getNotifyOthersError:
        getNotifyOthers = "0"
    End Function
```

Listing 5.6 The *Notification* **class of** *wfScriptLibrary*. **(continued)**

```
    '-----------------------------------------------------------------------


    ' getNotifyOriginator - retrieves the flag for specifying notification
        '                       of the originator
    Function getNotifyOriginator As String
        Dim i As NotesItem
        On Error Goto getNotifyOriginatorError


        Set i = curDoc.GetFirstItem("wfa_NotifyOriginator")
        getNotifyOriginator = i.Values(0)
        Exit Function


getNotifyOriginatorError:
        getNotifyOriginator = "0"
    End Function



    '-----------------------------------------------------------------------


    ' getNotifyStateOwner - retrieves the flag for specifying notification
        '                       of the state owner
    Function getNotifyStateOwner As String
        Dim i As NotesItem
        On Error Goto getNotifyStateOwnerError


        Set i = curDoc.GetFirstItem("wfa_NotifyOwner")
        getNotifyStateOwner = i.Values(0)
        Exit Function


getNotifyStateOwnerError:
        getNotifyStateOwner = "0"
    End Function
```

Listing 5.6 The *Notification* **class of** *wfScriptLibrary*. **(continued)**

```
'---------------------------------------------------------------------


    ' lookupAction - retrieves the action document
    Private Function lookupAction As NotesDocument
        Dim v As NotesView
        On Error Goto lookupActionError


        Set v = curDB.GetView("Workflow\wfActionsLookupByName")
        Set lookupAction = v.GetDocumentByKey(curKey)
        Exit Function


lookupActionError:
        Set lookupAction = Nothing
    End Function



    '---------------------------------------------------------------------


    ' notifyOthers - carry out the notification of others
    Sub notifyOthers
        Dim subject As String, text As Variant, actors As Variant
        Dim i As NotesItem
        On Error Goto notifyOthersError


            ' get the subject, text, and actors
        subject = sForm + " processing notice"
        Set i = curDoc.GetFirstItem("wfa_OtherText")
        text = i.Text
        If (Me.getNotifyOthersList = "") Then
            actors = Me.getNotifyOthersGroup
        Else
            actors = Me.getNotifyOthersList
        End If
```

Listing 5.6 The *Notification* **class of** *wfScriptLibrary*. **(continued)**

```
                   ' create the notification entry
             Call createNotification(subject, text, actors)
             Exit Sub


notifyOthersError:
             Exit Sub
        End Sub



             '----------------------------------------------------------------



             ' getNotifyOthersList - retrieve the notification list
             Function getNotifyOthersList As Variant
                Dim i As NotesItem
                On Error Goto getNotifyOthersListError


                Set i = curDoc.GetFirstItem("wfa_NotifyList")
                getNotifyOthersList = i.Text
                Exit Function


getNotifyOthersListError:
             Exit Function
        End Function



             '----------------------------------------------------------------



             ' getNotifyOthersGroup - retrieve the notification group
             Function getNotifyOthersGroup As Variant
                Dim i As NotesItem
                On Error Goto getNotifyOthersGroupError


                Set i = curDoc.GetFirstItem("wfa_NotifyOthersGroup")
                getNotifyOthersGroup = i.Text
                Exit Function
```

Listing 5.6 The *Notification* **class of** *wfScriptLibrary*. **(continued)**

```
getNotifyOthersGroupError:
    Exit Function
  End Function


    '-----------------------------------------------------------------
    ' notifyStateOwner - carry out the notification of the state owner
    Sub notifyStateOwner
       Dim subject As String, text As Variant, actors As Variant
       Dim i As NotesItem
       On Error Goto notifyStateOwnerError


          ' get the subject, text, and actors
       subject = sForm + " processing notice"
       Set i = curDoc.GetFirstItem("wfa_OwnerText")
       text = i.Text
       actors = sOwner


       ' create the notification entry
       Call createNotification(subject, text, actors)
       Exit Sub


notifyStateOwnerError:
    Exit Sub
  End Sub


    '-----------------------------------------------------------------
    ' notifyOriginator - carry out the notification of the originator
    Sub notifyOriginator
       Dim subject As String, text As Variant, actors As Variant
       Dim i As NotesItem
       On Error Goto notifyOriginatorError
```

Listing 5.6 The *Notification* **class of** *wfScriptLibrary*. **(continued)**

```
        ' get the subject, text, and actors
      subject = sForm + " processing notice"
      Set i = curDoc.GetFirstItem("wfa_OriginatorText")
      text = i.Text
      actors = sOriginator

       ' create the notification entry
      Call createNotification(subject, text, actors)
      Exit Sub

notifyOriginatorError:
      Exit Sub
   End Sub


   '-------------------------------------------------------------------


   ' createNotification - create a document in the notification engine
   Sub createNotification(subject As String, text As Variant, _
   actors As Variant)
      Dim db As NotesDatabase, doc As NotesDocument
      On Error Goto createNotificationError

         ' get the notification database and create a new document
      Set db = New NotesDatabase("", Me.getNotifyDB)
      Set doc = New NotesDocument(db)

         ' fill out the notification entry fields
      doc.Form = "Notification Entry"
      doc.RequestType = "1"
      doc.Requestor = "Workflow Management System"
      doc.Subject = subject
      doc.NotificationFrequency = "1"
```

Listing 5.6 The *Notification* **class of** *wfScriptLibrary.* **(continued)**

```
        doc.EntryBody = text
        doc.Actors = actors


            ' save the notification entry
        Call doc.ComputeWithForm(False, False)
        Call doc.Save(True, True)
        Exit Sub


createNotificationError:
        Exit Sub
    End Sub



    '----------------------------------------------------------------------



    ' getNotifyDB - retrieves the notification engine database
    Function getNotifyDB As String
        Dim v As NotesView, doc As NotesDocument, i As NotesItem
        On Error Goto getNotifyDBError


        Set v = curDB.GetView("Workflow\Profiles")
        Set doc = v.GetFirstDocument
        Set i = doc.GetFirstItem("wfp_NotificationDB")
        getNotifyDB = i.Text
        Exit Function


getNotifyDBError:
        getNotifyDB = "notify.nsf"
    End Function
End Class
```

5.7 Constructing the Rule Engine

The final piece of our States/Actions/Roles/Rules framework is the Rule Engine. An engine to process business rules can be simple or complex. An engine can implement a full production rule environment, such as in the design of many expert systems[1-4]. Or, the engine can be designed to handle a small number of rules that are used to make automated decisions.

The type of rules we will support are called production rules. These are of the form "If x Then y", where x is an antecedent clause that is either true or false. If the antecedent clause is evaluated and found to be true, the consequent clause y is executed.

For the purposes of this framework, we implement a relatively simple rule engine. The engine will deal with a limited set of rules, all of which are designed to execute in sequence. This is quite different from the design of forward-chaining expert systems, where rules are selected based on whether their antecedent clauses are true. All of these matching rules (in which the antecedents match the current environment, or set of conditions) are placed in a conflict set. Some method, differing among each expert system, is then used to choose one of the rules in the conflict set. This rule is executed, and the process is restarted, completing when the conflict set is empty.

The reason we want the workflow rule engine to deal with a limited rule set is two-fold: to process the rules in a specific order, and to make multiple decisions at the transition point between states. In this section, we present the complete source code for the simple rule engine, allowing you to extend and customize the engine for your own needs. You might, for example, wish to extend the engine to support multiple antecedent and multiple consequent clauses in each rule.

5.7.1 Representing the Rules

The first item we need to decide before commencing with the design and implementation of the rule engine is the format of the rules. In other words, how are we going to specify the rules? We need a way to create the rules easily, and to specify initial conditions, or facts, that can be used to test the antecedents of the rules for their truth values. In addition, the representation for the facts and rules must be easily associated to workflow actions. Finally, we need a way to terminate execution of a series of ordered rules that are associated with an action. We need to terminate execution if a particular rule causes a state transition. For example, in Figure 5.5C, Action 1 has two rules. If Rule 1 fires and causes a state transition to State 3, the rule engine should stop rather than continue testing the remaining rules.

Our choice for representing the logical operations associated with workflow business rules is XML. XML is readily created manually through a simple text editor, or generated by programs based on a graphical user interface. We can also take advantage of standard parsing tools to read and process XML data.

XML gives us the freedom to pick our own tags to denote the components of each fact and rule that describe the workflow logic. Consequently, we define an all-encompassing tag named WORKFLOWLOGIC to hold the FACTS and RULES. Each FACT tag includes a NAME and a VALUE. Each RULE tag contains a CONDITION (the If, or antecedent clause) and an ACTION (the Then, or consequent clause).

For example, consider the following three facts and rule that might be used to describe part of the workflow process depicted in Figure 5.1.

- Amount is a Notes field
- State is a Notes field
- limit is a variable with a value of 50
- If (Amount <= limit) Then (State = "Pending Payment")

The first two facts state that Amount and State are Notes fields that will be found in a Notes document being processed. The third fact just declares that limit is a constant with a value of 50.

The rule models the first decision point of Figure 5.1, at which the document's expense amount is compared to the constant of 50 dollars. If the condition is true, the document's state is automatically set to "Pending Payment".

These four items are encoded in an XML string for our framework as follows.

```xml
<?xml version="1.0" standalone="yes"?>
<WORKFLOWLOGIC>
    <FACTS>
        <NAME>Amount</NAME>
        <VALUE>NotesField</VALUE>
    </FACT>
    <FACT>
        <NAME>State</NAME>
        <VALUE>NotesField</VALUE>
    </FACT>
```

```
        <FACT>
           <NAME>limit</NAME>
           <VALUE>50</VALUE>
        </FACT>
    </FACTS>
    <RULES>
       <RULE TRANSITION="true">
          <CONDITION>
             <LHS>Amount</LHS>
             <OP>LESSTHANOREQUALTO</OP>
             <RHS>limit</RHS>
          </CONDITION>
          <ACTION>
             <LHS>State</LHS>
             <OP>EQUALTO</OP>
             <RHS>Pending Payment</RHS>
          </ACTION>
       </RULE>
    </RULES>
</WORKFLOWLOGIC>
```

Note that the <RULE> tag includes a TRANSITION attribute. If this attribute is "true", the rule is recognized as one with a consequent clause that causes a state transition. Thus, if the rule fires, the rule engine will carry out the state transition and stop.

5.7.2 Parsing the XML Rules

To read the XML representation of our rules into a Java agent for processing, we need a parser. The XML4J parser by IBM, available at the alphaWorks web site[5], is a leading XML parser engine.

Our rule engine will be based on a SAX (Simple API for XML) parser. This is an event-driven parser that is simple to use through implementation of a couple of Java interfaces. It is advantageous especially for parsing large XML documents, because the entire document does not have to be read at once.

As shown in Figure 5.12, the main class for the rule engine is called *wfLogic*. This class implements two interfaces from the XML4J package:

org.xml.sax.DocumentHandler and *org.xml.sax.ErrorHandler.* These interfaces provide methods that denote the start and end of parsing for a document, starting and ending of each tag, and methods to deliver the tag contents, any tag attributes, and error detection.

Figure 5.12 The Rule Engine package static structure diagram.

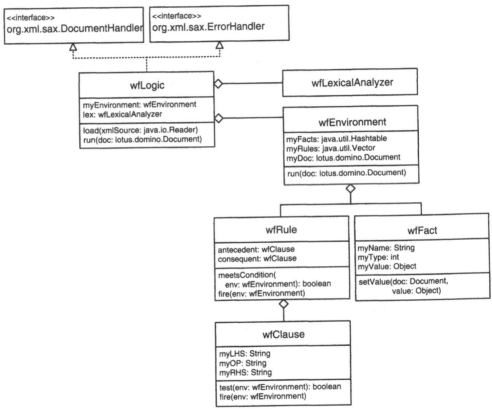

The *wfLogic* class maintains an instance of *wfLexicalAnalyzer*, whose methods are called during the event-driven parsing to detect each fact and rule. These facts and rules are returned to the *wfLogic* instance and placed into an instance of *wfEnvironment*.

5.7.3 Applying the Workflow Logic Rules

To apply the workflow logical operations on a particular document, the *wfLogic run()* method is called, passing in the *lotus.domino.Document* instance. This method in turn calls the *run()* method in *wfEnvironment*. The document becomes part of the environment for the evaluation of the antecedent clauses of the rules.

The logic evaluation proceeds by evaluating the antecedent clauses of the rules through the *meetsCondition()* method of *wfRule*. This method calls the *test()* method of the *wfClause* instance that represents the antecedent clause. During evaluation of the clause, various *wfFact* instances in the environment are used to return the values of variables, and compare values based on a specified operation (<, <=, =, !=, >=, >). For those antecedent clauses that return true, the *fire()* method is called to carry out the action specified in the consequent clause of the rule. This consequent might include setting the value of a Notes field in the document to a new value. This is done by calling the *setValue()* method of *wfFact*, passing it the document instance.

5.7.4 Implementing the Rule Engine Classes

Listing 5.7 shows the declaration of *wfLogic*, implementing the two XML4J interfaces described previously. There are two constructors: the first has an empty argument list, and the second has a *java.io.Reader* argument. This reader holds the XML source that is passed to a *load()* method. The *load()* method creates an instance of *wfEnvironment*, and calls an internal private method named *parseXML()* to parse the XML source.

Listing 5.7 The main properties and methods of the *wfLogic* **class.**

```
import java.util.*;
import java.io.*;
import com.ibm.xml.parsers.*;
import org.xml.sax.*;
import org.xml.sax.helpers.*;
import lotus.domino.*;
```

Listing 5.7 The main properties and methods of the *wfLogic* **class. (continued)**

```
public class wfLogic implements DocumentHandler, ErrorHandler
{
    // constructor
    public wfLogic()
    {
    }

    // ------------------------------------------------------------------

    // constructor
    public wfLogic(Reader xmlSource)
    {
        load(xmlSource);
    }

    // ------------------------------------------------------------------

    // load - parses an input XML specification of an execution environment
    //        and a set of rules that specify some logical workflow operations
    public void load(Reader xmlSource)
    {
        // set up the error flags
        logicLoaded = false;
        receivedWarnings = false;
        receivedErrors = false;
        receivedFatalErrors = false;

        // initialize the execution environment
        myEnvironment = new wfEnvironment();
```

Listing 5.7 The main properties and methods of the *wfLogic*
class. (continued)

```
    // parse the input
    parseXML(xmlSource);
}


// --------------------------------------------------------------------


// run - carries out the workflow logic operations (returns true if
//       the execution was completed, false if not)
public boolean run(Document doc)
{
    // make sure it is okay to run
    if (!logicLoaded) {
        notesLogFile.getInstance().write("wfLogic.run(): Not loaded");
        return false;
    }

    // run through the logic operations
    return myEnvironment.run(doc);
}


// --------------------------------------------------------------------


// anyErrors - returns true if any errors or warnings were detected when
//             the XML input was parsed
public boolean anyErrors()
{
    return (receivedWarnings || receivedErrors || receivedFatalErrors);
}
```

Listing 5.7 The main properties and methods of the *wfLogic* class. (continued)

```
// -------------------------------------------------------------------

// isUsable - returns true if the XML input was parsed and no
//            fatal errors were encountered
public boolean isUsable()
{
   return !receivedFatalErrors;
}


// -------------------------------------------------------------------


// toXML - returns an XML representation of the execution environment
public String toXML()
{
   // check to see if the environment exists
   if (myEnvironment == null) {
      // no environment - create a blank XML environment definition
      StringBuffer buf = new StringBuffer();
      buf.append("<?xml version=\"1.0\" standalone=\"yes\">\n");
      buf.append("<WORKFLOWLOGIC>\n");
      buf.append("   <FACTS></FACTS>\n   <RULES></RULES>\n");
      buf.append("</WORKFLOWLOGIC>\n");

      return new String(buf);
   } else
      return myEnvironment.toXML();
}

// -------------------------------------------------------------------
```

Listing 5.7 The main properties and methods of the *wfLogic* **class. (continued)**

```
// parseXML - parses XML input from a Reader
private void parseXML(Reader xmlSource)
{
   try {
      // make sure XML source is available
      if (xmlSource == null) {
         logicLoaded = false;
         return;
      }

      // create an XML parser
      Parser parser = ParserFactory.makeParser(DEFAULT_PARSER_NAME);
      parser.setDocumentHandler(this);
      parser.setErrorHandler(this);

      // parse the XML source
      parser.parse( new InputSource(xmlSource) );
      parser = null;

      // check for errors
      if ( isUsable() ) {
         logicLoaded = true;
      }
   }
   catch (Exception e) {
      notesLogFile.getInstance().write( "wfLogic.parseXML(): " +
                                 e.toString() );
   }
}
```

Listing 5.7 The main properties and methods of the *wfLogic*
class. (continued)

```
// **********************************************************************
// properties


// the facts and rules are encapsulated in an instance of wfEnvironment
private wfEnvironment myEnvironment = null;


// the following properties monitor the parsing of the XML input document
// that describes the execution environment
private boolean logicLoaded = false;


// lexical analyzer for creating tokens and rules during parsing
private wfLexicalAnalyzer lex = null;


// XML parser class
private static final String
    DEFAULT_PARSER_NAME = "com.ibm.xml.parsers.SAXParser";

}
```

The *parseXML()* method uses an *org.xml.sax.helpers.ParserFactory*
static method named *makeParser()* to create an instance of
com.ibm.xml.parsers.SAXParser. Then, the *wfLogic* instance is set as the
implementer of the *DocumentHandler* and *ErrorHandler* interfaces. The
parser is invoked to process the XML through a call to *parse()*.

In Listing 5.8, we implement the *DocumentHandler* interface with methods that the parser calls while processing the XML source. A *wfLexicalAnalyzer* instance is created and destroyed in response to *startDocument()*
and *endDocument()* events. As the parser encounters each new tag, the
startElement() method is called. This method in turn calls the lexical analyzer's *startTag()* method with the name of the tag. As the <RULE> tag

includes an attribute (TRANSITION), so the name and value of the attribute are also passed to *startTag()*.

Listing 5.8 The implementation of the *DocumentHandler* interface methods.

```
// DocumentHandler methods


// characters - receives notification of character data
public void characters(char[] ch, int start, int length)
{
    StringBuffer buf = new StringBuffer();
    for (int i = 0; i < length; i++)
        buf.append(ch[start + i]);

    lex.cdata( buf.toString() );
}


// -------------------------------------------------------------


// endDocument - receives notification of the end of a document
public void endDocument()
{
    // done with parsing
    lex = null;
}


// -------------------------------------------------------------


// endElement - receives notification of the end of an element
public void endElement(java.lang.String name)
{
    // if this ends the definition of a fact or rule, retrieve
    // it from the analyzer and place it in the environment
    lex.endTag(name);
```

Listing 5.8 The implementation of the *DocumentHandler* **interface methods. (continued)**

```
    if ( name.equals("FACT") && lex.hasFact() )
        myEnvironment.addFact( lex.getFact() );
    else if ( name.equals("RULE") && lex.hasRule() )
        myEnvironment.addRule( lex.getRule() );
}


// ----------------------------------------------------------------------


// ignorableWhitespace - receives notification of ignorable whitespace
//                       in element content
public void ignorableWhitespace(char[] ch, int start, int length)
{
}


// ----------------------------------------------------------------------


// processingInstruction - receives notification of a processing
//                         instruction
public void processingInstruction(java.lang.String target,
                                  java.lang.String data)

{
}


// ----------------------------------------------------------------------


// setDocumentLocator - receives an object for locating the origin of
//                      SAX document events
public void setDocumentLocator(Locator locator)

{
}


// ----------------------------------------------------------------------
```

Listing 5.8 The implementation of the *DocumentHandler* interface methods. (continued)

```
// startDocument - receives notification of the beginning of a document
public void startDocument()
{
    // create a new instance of the lexical analyzer
    lex = new wfLexicalAnalyzer();
}


// ----------------------------------------------------------------------


// startElement - receives notification of the beginning of an element
public void startElement(java.lang.String name, AttributeList atts)
{
    // if this is the start of a rule tag, get the value of the
    // transition attribute
    boolean transition = false;

    if ( name.equalsIgnoreCase("RULE") ) {
        // look at the attributes for the rule tag
        for (int i = 0; i < atts.getLength(); i++) {
            if ( atts.getName(i).equalsIgnoreCase("TRANSITION") &&
                 atts.getValue(i).equalsIgnoreCase("true") )
                transition = true;
        }
    }

    lex.startTag(name, transition);
}
```

The parser calls the *characters()* method to pass any character data that makes up the contents of a particular tag. We implement this method by forming a string from the array of characters, and pass the string to the lexical analyzer's *cdata()* method.

The *endElement()* event signals the end of processing for a given tag. This method calls the *endTag()* method of *wfLexicalAnalyzer* to finish processing of the tag. Then, if the processed tag is a FACT or a RULE, the corresponding *wfFact* or *wfRule* object is extracted from the analyzer and added to the execution environment.

The error handling for the parser can be done through exception handling, or through an implementation of the *ErrorHandler* interface (Listing 5.9). These methods simply set *boolean* properties to denote if any warnings, recoverable errors, or unrecoverable errors are encountered. The error messages are also written to a Notes log file through the *notesLogFile* class developed in Chapter 4.

Listing 5.9 Implementation of *ErrorHandler*.

```
// ErrorHandler interface methods


// error - receives notification of a recoverable error
public void error(SAXParseException exception)
{

   receivedErrors = true;
   notesLogFile.getInstance().write( "wfLogic.error(): " +
                                  exception.toString() );

}


// ----------------------------------------------------------------


// fatalError - receives notification of a non-recoverable error
public void fatalError(SAXParseException exception)
{

   receivedFatalErrors = true;
   notesLogFile.getInstance().write( "wfLogic.fatalError(): " +
                                  exception.toString() );

}


// ----------------------------------------------------------------
```

Listing 5.9 Implementation of *ErrorHandler*. (continued)

```
// warning - receives notification of a warning
public void warning(SAXParseException exception)
{
    receivedWarnings = true;
    notesLogFile.getInstance().write( "wfLogic.warning(): " +
                                      exception.toString() );
}


private boolean receivedWarnings = false;
private boolean receivedErrors = false;
private boolean receivedFatalErrors = false;
```

Once the parser is done, *wfLogic* includes two methods to check if the parsed XML is usable, or if errors were encountered. Another helper method in this class is *toXML()*. This method returns a string with the XML representation of the environment that was created during parsing.

To implement the *wfLexicalAnalyzer* class, the following properties are declared (Listing 5.10).

Listing 5.10 Instance and static properties of *wfLexicalAnalyzer*.

```
// properties


// temporaries used to form the tokens and rules
private String name = null;
private String value = null;
private String lhsCondition = null;       // for consequent clause
private String opCondition = null;
private String rhsCondition = null;
private String lhsAction = null;          // for antecedent clause
private String opAction = null;
private String rhsAction = null;
private boolean isTransition = false;   // value of rule TRANSITION attribute
private int expected = NOTFOUND;
private int clause = NOTFOUND;
```

Listing 5.10 Instance and static properties of *wfLexicalAnalyzer*. (continued)

```
// properties to return to parser
private wfFact curFact = null;
private wfRule curRule = null;

private boolean gotFact = false;    // determines when a complete fact
private boolean gotRule = false;    // or rule is defined

// for lookup operation on the tags
private static final int numTags = 12;
private static final String[] tags = {
    "WORKFLOWLOGIC", "FACTS", "FACT", "NAME",
    "VALUE", "RULES", "RULE", "CONDITION",
    "LHS", "OP", "RHS", "ACTION"
};

// integer values corresponding to tokens
private static final int WORKFLOWLOGIC = 1000;
private static final int FACTS         = 1001;
private static final int FACT          = 1002;
private static final int NAME          = 1003;
private static final int VALUE         = 1004;
private static final int RULES         = 1005;
private static final int RULE          = 1006;
private static final int CONDITION     = 1007;
private static final int LHS           = 1008;
private static final int OP            = 1009;
private static final int RHS           = 1010;
private static final int ACTION        = 1011;
private static final int NOTFOUND      = 1012;
```

The analyzer maintains some temporary strings to hold the name and value of the current *wfFact* being constructed, the left-hand-side, operation, and right-hand-side of the two clauses that make up the current *wfRule*. It also uses static variables to hold the strings corresponding to the legal tags recognized as part of the XML representation of the workflow logic source being parsed.

The work is done by the *startTag()*, *endTag()*, and *cdata()* methods shown in Listing 5.11. These methods call an internal method to convert the input string token into its corresponding integer value, and use a switch statement to process each token. As data is passed from the parser, the starting tag initializes to null the name, value, or other temporary properties appropriate to the tag. The contents of each XML tag are placed into the temporary variables in the *cdata()* method. Finally, a complete *wfFact* or *wfRule* instance is constructed by the *endTag()* method.

Listing 5.11 The *wfLexicalAnalyzer* methods called during parsing of XML source documents.

```
// startTag - called by parser when a new start tag is encountered
public void startTag(String token, boolean transition)
{
    // starting a tag invalidates any current fact or rule
    gotFact = false;
    gotRule = false;
    expected = NOTFOUND;

    // process the tag based on type
    int curType = lookup(token);

    switch (curType) {
        case WORKFLOWLOGIC:
        case FACTS:
        case RULES:
        case NOTFOUND:
            break;
```

Listing 5.11 The *wfLexicalAnalyzer* **methods called during parsing of XML source documents. (continued)**

```
    case RULE:
        isTransition = transition;
        break;

    case FACT:
        name = null;
        value = null;
        break;

    case NAME:
    case VALUE:
    case LHS:
    case OP:
    case RHS:
        expected = curType;
        break;

    case CONDITION:
        lhsCondition = null;
        opCondition = null;
        rhsCondition = null;
        clause = CONDITION;
        break;

    case ACTION:
        lhsAction = null;
        opAction = null;
        rhsAction = null;
        clause = ACTION;
        break;
    }
}
```

Listing 5.11 The *wfLexicalAnalyzer* **methods called during parsing of XML source documents. (continued)**

```
//  -----------------------------------------------------------------------


// endTag - called by parser when an end tag is encountered
public void endTag(String token)
{
    // create the token or rule
    switch ( lookup(token) ) {
        case WORKFLOWLOGIC:
        case FACTS:
        case NAME:
        case VALUE:
        case RULES:
        case CONDITION:
        case LHS:
        case OP:
        case RHS:
        case ACTION:
        case NOTFOUND:
            break;

        case FACT:
            curFact = new wfFact(name, value);
            gotFact = true;
            break;
```

Listing 5.11 The *wfLexicalAnalyzer* **methods called during parsing of XML source documents. (continued)**

```
        case RULE:
            curRule = new wfRule(
                new wfClause(lhsCondition, opCondition, rhsCondition),
                new wfClause(lhsAction, opAction, rhsAction), isTransition );
            gotRule = true;
            break;

    }


    expected = NOTFOUND;      // done with this tag
}


// -----------------------------------------------------------------------

// cdata - called by parser when character data is encountered
public void cdata(String token)
{
    // the type of data expected was set when the start tag was encountered
    switch (expected) {
        case WORKFLOWLOGIC:
        case FACTS:
        case FACT:
        case RULES:
        case RULE:
        case CONDITION:
        case ACTION:
        case NOTFOUND:
            break;

        case NAME:
            name = new String(token);
            break;
```

Listing 5.11 The *wfLexicalAnalyzer* **methods called during parsing of XML source documents. (continued)**

```
        case VALUE:
            value = new String(token);
            break;


        case LHS:
            if (clause == CONDITION)
                lhsCondition = new String(token);
            else if (clause == ACTION)
                lhsAction = new String(token);
            break;


        case OP:
            if (clause == CONDITION)
                opCondition = new String(token);
            else if (clause == ACTION)
                opAction = new String(token);
            break;


        case RHS:
            if (clause == CONDITION)
                rhsCondition = new String(token);
            else if (clause == ACTION)
                rhsAction = new String(token);
            break;
    }

}
```

Once the analyzer completes the creation of a fact or rule, it is added to the execution environment. This is done by calling the *addFact()* and *addRule()* methods of *wfEnvironment* (Listing 5.12). These methods add facts and rules to internal instances of *java.util.Hashtable* and *java.util.Vector*, respectively. The other methods in this listing test if a particular fact is in the environment, retrieve it, or delete it.

Listing 5.12 The *wfEnvironment* **methods and properties that contain the facts and rules of the workflow logic operations.**

```
// isDefined - returns true if the specified fact name
//              is in the facts table
public boolean isDefined(String factName)
{
    // make sure the specified symbol is valid
    if ( (factName == null) || (factName.length() == 0) )
        return false;

    return myFacts.containsKey(factName);
}


// --------------------------------------------------------------


// addFact - adds a fact to the environment
public void addFact(wfFact fact)
{
    // if the fact is valid, put it in the table
    if (fact != null)
        myFacts.put(fact.getName(), fact);
}


// --------------------------------------------------------------
```

Listing 5.12 The *wfEnvironment* **methods and properties that contain the facts and rules of the workflow logic operations. (continued)**

```
// getFact - retrieves a fact by name
public wfFact getFact(String factName)
{
    // make sure the fact is in the table
    if ( !isDefined(factName) )
        return null;

    return (wfFact)myFacts.get(factName);
}


// --------------------------------------------------------------------


// deleteFact - deletes a fact from the environment
public void deleteFact(String factName)
{
    // if the fact is in the table, delete it
    if ( isDefined(factName) )
        myFacts.remove(factName);
}


// --------------------------------------------------------------------


// addRule - adds a rule to the environment
public void addRule(wfRule rule)
{
    myRules.addElement(rule);
}
```

Listing 5.12 The *wfEnvironment* **methods and properties that contain the facts and rules of the workflow logic operations. (continued)**

```
// ********************************************************************
// properties


// facts and rules that define the environment
private Hashtable myFacts = null;
private Vector myRules = null;
```

To run the workflow logic on a particular document, the *wfLogic run()* method is called. This method simply calls the *run()* method of *wfEnvironment*, passing it the current document. This *Document* instance becomes part of the environment during the processing, as shown in Listing 5.13.

Listing 5.13 The *run()* **method of** *wfEnvironment.*

```
// run - runs the logic operations in this environment
public boolean run(Document doc)
{
    // if there is a document argument, save it as the current document
    myDoc = doc;


    // enumerate and execute the rules in the environment until all
    // rules have been fired, or a rule is fired that causes a state
    // change (note that this is different from an expert system,
    // where rules would be added to a conflict set and
    // a subset chosen for execution)
    for (int i = 0; i < myRules.size(); i++) {
        // get the next rule
        wfRule rule = (wfRule)myRules.elementAt(i);
        if (rule != null) {
            // if the rule meets its condition, execute the action
            if ( rule.meetsCondition(this) ) {
                rule.fire(this);

                // if this rule causes a state transition, stop
```

Listing 5.13 The *run()* method of *wfEnvironment*. (continued)

```
                    // the logic operations at this point
                    if ( rule.causesTransition() )
                        return true;

                }

            }

        }


    return true;

}
```

To execute the logic rules, the *run()* method simply iterates through each rule, calling the rule's *meetsCondition()* method. This tests if the antecedent clause of the rule is true. If so, the rule is fired. Once fired, the rule is tested to see if it should be the last one processed. If so, the *run()* method returns.

The testing and firing methods of the rule are shown in the listing of the complete *wfRule* class (Listing 5.14). This class holds two instances of *wfClause*, one each for the antecedent and consequent portions of the rule. The testing and firing methods call the appropriate methods, *test()* and *fire()*, of *wfClause*.

Listing 5.14 The *wfRule* class.

```
// ***************************************************************************
//
// Module: wfRule.java
// Author: Dick Lam
//
// Description: This is a Java class that holds an If...Then rule. It
//              includes methods to test whether the If clause is true,
//              and to fire the rule by carrying out the operation
//              specified in the Then clause.
//
// ***************************************************************************

import java.io.*;
```

Listing 5.14 The *wfRule* **class. (continued)**

```
// *****************************************************************************

public class wfRule
{
  // constructor
  public wfRule(wfClause ifClause, wfClause thenClause, boolean transition)
  {
     antecedent = ifClause;
     consequent = thenClause;
     isTransition = transition;
  }

  // --------------------------------------------------------------------

  // getCondition - returns the condition clause
  public wfClause getCondition()
  {
     return antecedent;
  }

  // --------------------------------------------------------------------

  // getAction - returns the consequent clause
  public wfClause getAction()
  {
     return consequent;
  }
```

Listing 5.14 The *wfRule* **class. (continued)**

```
// --------------------------------------------------------------------------

// meetsCondition - tests the 'If' clause
public boolean meetsCondition(wfEnvironment env)
{
   if (antecedent != null)
     return antecedent.test(env);
   else
     return false;
}

// --------------------------------------------------------------------------

// fire - fires the 'Then' clause
public void fire(wfEnvironment env)
{
   if (consequent != null)
     consequent.fire(env);
}

// --------------------------------------------------------------------------

// causesTransition - returns true if the rule causes a state
//                    transition when fired
public boolean causesTransition()
{
   return isTransition;
}
```

Listing 5.14 The *wfRule* class. (continued)

```
// ***********************************************************************
// properties


private wfClause antecedent = null;        // the 'If' clause
private wfClause consequent = null;        // the 'Then' clause
private boolean isTransition = false;      // true if this rule will
                                           // cause a state transition

}


// ***********************************************************************


// end of wfRule.java
```

The *test()* method of *wfClause* is shown in Listing 5.15, along with the internal support methods. Note that the current environment is passed to this method, and temporarily saved as a private property of the clause. To start the test of the antecedent clause, the *evaluate()* method is called for both the left-hand side and right-hand side of the clause. This returns two *wfFact* objects that can be compared using the operation associated with the clause.

Listing 5.15 The methods involved in the *test()* method of *wfClause*.

```
// test - evaluates the clause and returns true or false
public boolean test(wfEnvironment env)
{
    // make sure environment is valid
    if (env == null)
        return false;

    // evaluate the lhs and rhs
    curEnv = env;
    wfFact t1 = evaluate(myLHS);
    wfFact t2 = evaluate(myRHS);
```

Listing 5.15 The methods involved in the *test()* **method of** *wfClause*. **(continued)**

```
// carry out the operation and return the result
switch ( lookupOperation(myOP) ) {
    case LESSTHANOREQUALTO:
        return lessThanOrEqualTo(t1, t2);

    case LESSTHAN:
        return lessThan(t1, t2);

    case EQUALTO:
        return equalTo(t1, t2);

    case NOTEQUALTO:
        return !equalTo(t1, t2);

    case GREATERTHAN:
        return !lessThanOrEqualTo(t1, t2);

    case GREATERTHANOREQUALTO:
        return !lessThan(t1, t2);

    default:
        return false;
    }
}

// ----------------------------------------------------------------
```

Listing 5.15 The methods involved in the `test()` **method of**
`wfClause`. **(continued)**

```
// evaluate - evaluates the input argument
private wfFact evaluate(String s)
{
   // if the input argument is not part of the fact list of the
   // environment, just return a symbol
   if ( !curEnv.isDefined(s) )
      return new wfFact(s, s);

   // get the matching fact from the environment
   wfFact fact = curEnv.getFact(s);
   if (fact.getType() != wfFact.SYMBOL)
      return fact;

   // get the symbol value and evaluate it
   String value = (String)fact.getValue(null);
   if ( curEnv.isDefined(s) )
      return evaluate(value);
   else
      return fact;
}


// -----------------------------------------------------------------------

// lessThanOrEqualTo - compares the values of the two input arguments
private boolean lessThanOrEqualTo(wfFact t1, wfFact t2)
{
   if ( lessThan(t1, t2) )
      return true;
   if ( equalTo(t1, t2) )
      return true;
```

Listing 5.15 The methods involved in the `test()` **method of** `wfClause`. **(continued)**

```
    return false;
}


// --------------------------------------------------------------


// lessThan - compares the values of the two input arguments
private boolean lessThan(wfFact t1, wfFact t2)
{
    // if either argument is a Notes field, evaluate it
    if ( t1.isNotesField() )
        t1.getValue( curEnv.getDocument() );
    if ( t2.isNotesField() )
        t2.getValue( curEnv.getDocument() );

    return t1.lessThan(t2);
}


// --------------------------------------------------------------


// equalTo - compares the values of the two input arguments
private boolean equalTo(wfFact t1, wfFact t2)
{
    // if either argument is a Notes field, evaluate it
    if ( t1.isNotesField() )
        t1.getValue( curEnv.getDocument() );
    if ( t2.isNotesField() )
        t2.getValue( curEnv.getDocument() );

    return t1.equalTo(t2);
}
```

All of the operations (i.e., less than, less than or equal to, equal to, not equal, greater than or equal to, and greater than) can be performed using the two private methods *lessThan()* and *equalTo()*. For example, if the two facts do not have a "less than" relationship, their relationship must be "greater than or equal to". Both of the comparison methods work by calling the corresponding method defined in the *wfFact* class.

The *evaluate()* method uses the current environment to determine the value of the input fact before the comparison begins. The first check in *evaluate()* determines if the symbol is defined in the environment. If not, the symbol is simply returned. Otherwise, the fact is retrieved from the environment. If the fact is a symbol, its value is retrieved and recursively evaluated. This lets the rule engine handle variables that refer to their value as another variable (e.g., limit = x, and x = 50, so limit will evaluate to 50).

The last class we must implement to complete the rule engine is *wfFact*. Each fact has a name, a type, and a value. For the purposes of this framework, we let the possible types of values be either a *Long*, a *Double*, a *String*, or a Notes field name. The constructors of this class call the internal method *setType()* to determine the type associated with the name, and to store the value of a *Long*, *Double*, or *String* type.

The value of a fact is returned by the *getValue()* method, which takes an optional argument of a Notes document instance. For those facts that represent Notes fields, the value is retrieved directly from the Notes document, using the *getFieldValue()* method (see Listing 5.16).

Listing 5.16 The *wfFact* class.

```
// *****************************************************************************
//
// Module: wfFact.java
// Author: Dick Lam
//
// Description: This class defines facts (variables and their associated
//              values) that are used for workflow logic evaluation.
//
// *****************************************************************************


import java.util.*;
import lotus.domino.*;


// *****************************************************************************
```

Listing 5.16 The *wfFact* class. (continued)

```
public class wfFact
{
    // constructor
    public wfFact(String name, String value)
    {
        // make sure name and value are valid
        if ( (name == null) || (name.length() == 0) ||
             (value == null) || (value.length() == 0) )
            return;

        // save the name and value, and determine the type of the value
        myName = new String(name);
        setType(value);
    }

    // -----------------------------------------------------------------

    // constructor
    public wfFact(String name, Object value)
    {
        // make sure name and value are valid
        if ( (name == null) || (name.length() == 0) || (value == null) )
            return;

        // save the name and value, and determine the type of the value
        myName = new String(name);
        setType( value.toString() );
    }

    // -----------------------------------------------------------------
```

Listing 5.16 The *wfFact* class. (continued)

```
// getName - returns the name of the variable or Notes field
public String getName()
{
   return new String(myName);
}


// ----------------------------------------------------------------


// getType - returns the type of the value (i.e., one of the types
//           listed in the public static area)
public int getType()
{
   return myType;
}


// ----------------------------------------------------------------


// getValue - returns the value (the doc argument may be null for
//            all cases except when the value type is NOTES_FIELD)
public Object getValue(Document doc)
{
   // return an object based on the value type
   switch (myType) {
      case NUMERIC_LONG:
         return new Long(myLongValue);

      case NUMERIC_DOUBLE:
         return new Double(myDoubleValue);

      case SYMBOL:
         return new String(mySymbolValue);
```

Listing 5.16 The *wfFact* class. (continued)

```
        case NOTES_FIELD:
            return getFieldValue(doc);


        default:
            return null;
    }
}


// ------------------------------------------------------------------------


// setValue - sets the value of this instance (if the instance is
//            of type NOTES_FIELD) to the specified value
public void setValue(Document doc, Object value)
{
    try {
        // make sure this instance is a Notes field and the argument
        // is valid
        if ( (myType != NOTES_FIELD) || (doc == null) || (value == null) )
            return;


        // set the value
        doc.replaceItemValue(myName, value);
        doc.save(true);
    }
    catch (Exception e) {
        notesLogFile.getInstance().write( "wfFact.setValue(): " +
                                    e.toString() );
    }
}


// ------------------------------------------------------------------------
```

Listing 5.16 The *wfFact* class. (continued)

```java
// isSymbol - returns true if the token represents a symbol
public boolean isSymbol()
{
    return (myType == SYMBOL);
}



// ------------------------------------------------------------



// isNumber - returns true if the token is a number
public boolean isNumber()
{
    return ( (myType == NUMERIC_LONG) || (myType == NUMERIC_DOUBLE) );
}



// ------------------------------------------------------------



// isNotesField - returns true if the token is the name of a Notes field
public boolean isNotesField()
{
    return (myType == NOTES_FIELD);
}



// ------------------------------------------------------------



// lessThan - test if the current value is less than the argument value
public boolean lessThan(wfFact f)
{
    // compare based on type of value
    switch ( f.getType() ) {
      case NUMERIC_LONG:
        return lessThan(f.myLongValue);
```

Listing 5.16 The *wfFact* class. (continued)

```
        case NUMERIC_DOUBLE:
            return lessThan(f.myDoubleValue);

        case SYMBOL:
            return lessThan(f.mySymbolValue);

        case NOTES_FIELD:
            return lessThan(f.myNotesValue);

        default:
            return false;
    }
}

// -------------------------------------------------------------------

// equalTo - test if the current value is equal to the argument value
public boolean equalTo(wfFact f)
{
    // compare based on type of value
    switch ( f.getType() ) {
        case NUMERIC_LONG:
            return equalTo(f.myLongValue);

        case NUMERIC_DOUBLE:
            return equalTo(f.myDoubleValue);

        case SYMBOL:
            return equalTo(f.mySymbolValue);

        case NOTES_FIELD:
            return equalTo(f.myNotesValue);
```

Listing 5.16 The *wfFact* **class. (continued)**

```
        default:
            return false;

    }

}


// --------------------------------------------------------------------


// lessThan - test if the current value is less than the argument value
private boolean lessThan(long l)
{
    switch (myType) {
        case NUMERIC_LONG:
            return (myLongValue < l);


        case NUMERIC_DOUBLE:
            Double d = new Double(l);
            return ( myDoubleValue < d.doubleValue() );


        case SYMBOL:
            return false;


        case NOTES_FIELD:
            if ( (myNotesValue == null) || (myNotesValue.size() == 0) )
                return false;
            wfFact f = new wfFact( "f", myNotesValue.elementAt(0) );
            return f.lessThan(l);


        default:
            return false;

    }

}
```

Listing 5.16 The *wfFact* **class. (continued)**

```
// ---------------------------------------------------------------

// lessThan - test if the current value is less than the argument value
private boolean lessThan(double d)
{
    switch (myType) {
        case NUMERIC_LONG:
            Double x = new Double(myLongValue);
            return (x.doubleValue() < d);

        case NUMERIC_DOUBLE:
            return (myDoubleValue < d);

        case SYMBOL:
            return false;

        case NOTES_FIELD:
            if ( (myNotesValue == null) || (myNotesValue.size() == 0) )
                return false;
            wfFact f = new wfFact( "f", myNotesValue.elementAt(0) );
            return f.lessThan(d);

        default:
            return false;
    }
}

// ---------------------------------------------------------------
```

Listing 5.16 The *wfFact* class. (continued)

```
// lessThan - test if the current value is less than the argument value
private boolean lessThan(String s)
{
    switch (myType) {
        case NUMERIC_LONG:
        case NUMERIC_DOUBLE:
            return false;

        case SYMBOL:
            return (mySymbolValue.compareTo(s) < 0);

        case NOTES_FIELD:
            if ( (myNotesValue == null) || (myNotesValue.size() == 0) )
                return false;
            wfFact f = new wfFact( "f", myNotesValue.elementAt(0) );
            return f.lessThan(s);

        default:
            return false;
    }
}

// ----------------------------------------------------------------------

// lessThan - test if the current value is less than the argument value
private boolean lessThan(Vector v)
{
    // see if there is anything in the input vector
    if ( (v == null) || (v.size() == 0) )
        return false;

    switch (myType) {
        case NUMERIC_LONG:
```

Listing 5.16 The *wfFact* class. (continued)

```
        case NUMERIC_DOUBLE:
        case SYMBOL:
            // compare the value to the first vector element
            return lessThan( new wfFact( "f", v.elementAt(0) ) );

        case NOTES_FIELD:
            // compare all elements of the two vectors
            if ( (myNotesValue == null) || (myNotesValue.size() == 0) )
                return false;
            int num = myNotesValue.size();
            if ( num != v.size() )
                return false;
            boolean status = true;
            for (int i = 0; i < v.size(); i++) {
                wfFact f = new wfFact( "f", myNotesValue.elementAt(i) );
                status &= f.lessThan( new wfFact( "f", v.elementAt(i) ) );
                f = null;
            }
            return status;


        default:
            return false;
    }
}


// --------------------------------------------------------------------

// equalTo - test if the current value is equal to the argument value
private boolean equalTo(long l)
{
    switch (myType) {
        case NUMERIC_LONG:
            return (myLongValue == l);
```

Listing 5.16 The *wfFact* **class. (continued)**

```
        case NUMERIC_DOUBLE:
           Double d = new Double(1);
           return ( myDoubleValue == d.doubleValue() );


        case SYMBOL:
           return false;

        case NOTES_FIELD:
           if ( (myNotesValue == null) || (myNotesValue.size() == 0) )
              return false;
           wfFact f = new wfFact( "f", myNotesValue.elementAt(0) );
           return f.equalTo(1);


        default:
           return false;
      }
   }


   // -----------------------------------------------------------------


   // equalTo - test if the current value is equal to the argument value
   private boolean equalTo(double d)
   {
      switch (myType) {
        case NUMERIC_LONG:
           Double x = new Double(myLongValue);
           return (x.doubleValue() == d);

        case NUMERIC_DOUBLE:
           return (myDoubleValue == d);
```

Listing 5.16 The *wfFact* **class. (continued)**

```
        case SYMBOL:
          return false;


        case NOTES_FIELD:
          if ( (myNotesValue == null) || (myNotesValue.size() == 0) )
            return false;
          wfFact f = new wfFact( "f", myNotesValue.elementAt(0) );
          return f.equalTo(d);


        default:
          return false;
      }
    }


    // ---------------------------------------------------------------------


    // equalTo - test if the current value is equal to the argument value
    private boolean equalTo(String s)
    {
      switch (myType) {
        case NUMERIC_LONG:
        case NUMERIC_DOUBLE:
          return false;


        case SYMBOL:
          return mySymbolValue.equalsIgnoreCase(s);


        case NOTES_FIELD:
          if ( (myNotesValue == null) || (myNotesValue.size() == 0) )
            return false;
          wfFact f = new wfFact( "f", myNotesValue.elementAt(0) );
          return f.equalTo(s);
```

Listing 5.16 The *wfFact* class. (continued)

```
        default:
            return false;
    }
}


// ----------------------------------------------------------------------


// equalTo - test if the current value is equal to the argument value
private boolean equalTo(Vector v)
{
    // see if there is anything in the input vector
    if ( (v == null) || (v.size() == 0) )
        return false;

    switch (myType) {
        case NUMERIC_LONG:
        case NUMERIC_DOUBLE:
        case SYMBOL:
            // compare the value to the first vector element
            return equalTo( new wfFact( "f", v.elementAt(0) ) );

        case NOTES_FIELD:
            // compare all elements of the two vectors
            if ( (myNotesValue == null) || (myNotesValue.size() == 0) )
                return false;
            int num = myNotesValue.size();
            if ( num != v.size() )
                return false;
            boolean status = true;
```

Listing 5.16 The *wfFact* **class. (continued)**

```
        for (int i = 0; i < v.size(); i++) {
            wfFact f = new wfFact( "f", myNotesValue.elementAt(i) );
            status &= f.equalTo( new wfFact( "f", v.elementAt(i) ) );
            f = null;
        }
        return status;

    default:
        return false;
    }
}


// --------------------------------------------------------------------

// this method determines the type of the current token
private void setType(String value)
{
    // check for the object type represented by value
    if ( value.equalsIgnoreCase("NotesField") )
        myType = NOTES_FIELD;
    else if ( isLong(value) )
        myType = NUMERIC_LONG;
    else if ( isDouble(value) )
        myType = NUMERIC_DOUBLE;
    else {
        // must be a symbol
        mySymbolValue = new String(value);
        myType = SYMBOL;
    }
}
```

Listing 5.16 The *wfFact* **class. (continued)**

```java
// ------------------------------------------------------------------

// isLong - checks if the current token is a parsable Long
private boolean isLong(String value)
{
   try {
      myLongValue = Long.parseLong(value);
   }
   catch (NumberFormatException e) {
      return false;
   }

   return true;
}


// ------------------------------------------------------------------

// isDouble - checks if the current token is a parsable Double
private boolean isDouble(String value)
{
   try {
      Double d = Double.valueOf(value);
      myDoubleValue = d.doubleValue();
   }
   catch (NumberFormatException e) {
      return false;
   }

   return true;
}
```

Listing 5.16 The *wfFact* class. (continued)

```
// ----------------------------------------------------------------------


// getFieldValue - returns the value of a Notes field
private Object getFieldValue(Document doc)
{
  try {
    // make sure the argument is valid
    if ( (doc == null) || !doc.hasItem(myName) )
      return null;


    myNotesValue = doc.getItemValue(myName);
    return myNotesValue;
  }
  catch (Exception e) {
    return null;
  }
}


// ************************************************************************
// properties

private String myName = null;          // name
private int myType = UNSPECIFIED;       // type


// value
private long myLongValue = 0;
private double myDoubleValue = (double)0;
private String mySymbolValue = null;
private Vector myNotesValue = null;
```

Listing 5.16 The *wfFact* **class. (continued)**

```
    // these integers denote specific token types
    static public final int UNSPECIFIED    = 1000;
    static public final int NUMERIC_LONG    = 1001;
    static public final int NUMERIC_DOUBLE = 1002;
    static public final int SYMBOL          = 1003;
    static public final int NOTES_FIELD     = 1004;
}


// **********************************************************************


// end of wfFact.java
```

The other methods in this class are to facilitate comparisons based on the *lessThan()* and *equalTo()* methods. There are versions of these methods for each of the four types of values.

Note that it is possible to extend this class to hold other value types (e.g., Notes *DateTime* values). However, a better object-oriented design would separate the various value types into derived classes of *wfFact*. Then, new value types would be supported by adding additional derived classes and specializing the value property and the comparison methods.

5.8 Summary

This has been a long chapter. The journey, however, has brought us through the design and implementation of a usable and extensible workflow management system. Basing our design on Notes/Domino enables the framework to be used in both new and existing Domino applications. The framework is designed explicitly for Web-based systems accessible through a browser interface. It also allows configuration of a workflow process through a set of Notes forms, and definition of business rules using XML to facilitate automated workflow decisions.

In the next two chapters, we finish the book by exploring the use of the framework through a case study. In particular, we focus on a sample college admissions office, and model some of the workflow processes that support the admissions office functions.

5.9 References

1   J.P. Bigus and J. Bigus. 1998. *Constructing Intelligent Agents with Java.* New York, NY: John Wiley & Sons.

2   Mark Watson. 1997. *Intelligent Java Applications for the Internet and Intranets.* San Francisco, CA: Morgan Kaufman Publishers.

3   S.J. Metsker. February, 1998. "Java Rules: Develop Powerful Interpreters with Java", *Java Report.*

4   D. Phillips. February, 1998. "Decision-Making with Production Systems", *C/C++ Users Journal.*

5   IBM alphaWorks. [http://www.alphaworks.ibm.com]

Chapter 6

A Case Study: College Admissions Processing

In the next two chapters, we show how to apply the framework we have built. To demonstrate the application of the framework to a "real-world" problem, we present a case study. Our topic is something familiar to most readers: applying to college.

In this chapter, we start by examining a fictional university. The admissions process for most colleges and universities are basically similar. We first outline the current paper-based approaches to admissions procedures. Next, we gather requirements, analyze the processes, and design some initial forms for various stages of our model.

The last chapter picks up where this one leaves off. We take the preliminary design specifications, forms, and views already designed, and workflow-enable a selected process. The result is an online admissions processing system that demonstrates how to use the workflow framework.

Note that we do not intend the final result to be a complete workflow model for a real university admissions process. However, it should provide you with a starting point in how to apply this book's workflow package in genuine customer engagements. The intent is to provide a life-like model of a familiar process that differs from ones typically talked about in traditional workflow books and papers.

6.1 The Interview

Let's start by visiting our fictional campus: Wyrcan Flowan University. We meet with the admissions office director, Kelly. Kelly has hired consultants to reengineer the university admissions process. The university has funded an effort to create a total online "e-admissions" system. Because the IT department already uses Lotus Notes campus-wide for e-mail and groupware purposes, it has decided that the new system will be based on Domino.

The first task is to examine the overall department structure of the university so we can see where the admissions office fits in with the other university offices. This will provide information on where potential workflow processes might require involvement of other offices. This is important, as each of these offices may have their own well-defined processes, and may handle or store documents in other types of systems.

Kelly describes the different offices that make up the university. Figure 6.1 shows the Admissions office, and all of the other relevant offices in the university with which Admissions typically collaborates. Note that Admissions differs from other departments in that it must work closely with many offices. It requires input from Academic Programs, the Bursar, and Administration to make admission decisions. It must also interface with Housing, the Registrar, and Financial Aid to provide information to accepted applicants.

The second task is to get an introduction to the admissions staff. Kelly explains that she is the office director. She manages a staff of four assistant directors, five associate directors, and five support staff. The assistant and associate directors make up the professional staff. These people interface with three registrar office personnel and five academic department heads. The professional staff also includes a director for a summer developmental program.

Next, we need to get a handle on the general approach to admissions, and the scale of various numbers associated with the admissions process. In Wyrcan Flowan University, we learn that graduate and undergraduate admissions are handled separately, and the current consulting contract is to reengineer undergraduate admissions of entering freshman only. A follow-on contract may be let to handle transfer students.

Kelly describes the current admissions process. All applications are received on paper forms. Each application is handled individually through a holistic approach. The general criteria reviewed for each applicant are their high school class ranking, an essay, SAT scores, GPA, the strength of the high school curriculum, and the number of college preparatory courses taken.

Figure 6.1 The departments that interface with the Admissions office of a fictional university.

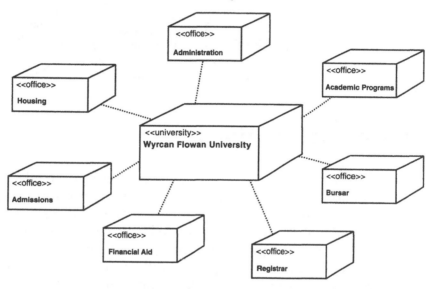

Finally, Kelly explains that the office receives nine thousand applications per year, and only twelve hundred freshman are accepted. She laments that up to three hundred applications may be received in one day. Given her limited staff and the number of applications they must review in detail, she wants to convert to a totally Web-based application process. She also outlines the three main processes that she and her staff carry out. These processes are shown in Figure 6.2.

Figure 6.2 The three major functions of the Admissions office.

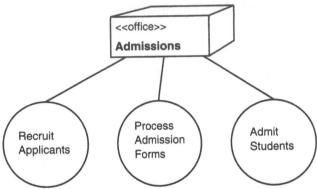

6.2 Case Assumptions

To simplify this case study problem, Wyrcan Flowan University is public, with no early decision program. Also, our first implementation of the e-admissions system does not have to deal with in-state versus out-of-state students, transfers, or admission of older students. Only graduating high school seniors are to be processed by the online system.

As seen in Figure 6.2, although recruiting is a major function of the office, the e-admissions process will be concerned with only two functions: processing online admissions forms and admitting students to the next freshman class.

A further simplifying assumption we make is that the e-commerce portion of the e-admissions process, namely payment of application fees, deposits, etc., is not implemented.

6.3 Communications-Based Models

Based on our interview with Kelly, we can look at the overall communications model of the admissions process. This model is shown in Figure 6.3. It is our standard workflow loop introduced in Chapter 2.

The customer, from the point of view of the admissions process, is the student applicant. The performer is the admissions office, where the decision to accept or reject an applicant is made.

The request phase of the workflow loop is initiated when the student applicant applies for admission to the university. One of our tasks in this study is to make the initial application process available through the Web.

Figure 6.3 A communications-based model of the admissions process.

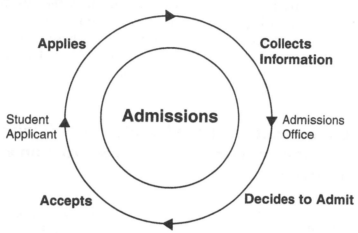

Once an application has been received, a number of different pieces of information must be collected by the Admissions office. These include the student transcripts, SAT scores, essay, etc. An applications fee must also be received to initiate review of the application. Another task in our e-admissions system should track what information is still required and notify or remind the appropriate parties.

The personnel in the Admissions office carry out the performance phase. This is the point in the process at which the admit/reject decision on each applicant is made. Other offices, particularly academic departments in the applicant's requested major, may be asked to review and provide recommendations at this point. The e-admissions system should be able to report on the status of each application during this phase. It should allow applications to be reviewed online by the professional staff, capturing comments to support the decision process.

Finally, the acceptance phase is the point at which the applicant is notified of the admission decision: acceptance or rejection. Acceptance of an applicant can initiate an entirely new workflow process. A positive admission decision requires the applicant to indicate their acceptance and intent to enroll for the next term. Thus, the follow-up processes mail information packets on housing options, collect deposits, schedule orientation visits, provide financial aid information, etc.

Figure 6.4 shows another communications-based model. This model describes the secondary workflow process that takes place during the performance phase of the Admissions process. Here, a student applicant has requested to enroll in a specific major or department (e.g., a pre-medical program).

In this model, the Admissions department is the customer. A potential student applicant is proposed for admission into a particular program. The academic department handles the negotiation phase by reviewing the application. An interview or audition may be carried out. The performance phase generates a recommendation to the Admissions office, which receives the information and continues to process the applicant's admission request.

Figure 6.4 A communications-based model for obtaining recommendations on admission from an academic department.

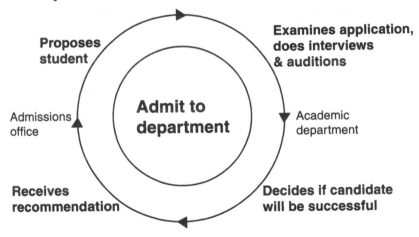

6.4 Process Models

In discussing our communications-based models, we alluded to a number of subprocesses related to the general admissions process. Let's consider the performance phase of Figure 6.3. Our admissions director described the subprocesses associated with the decision process as outlined in Figure 6.5.

One of the requirements for the e-admissions system is that all information is requested or collected before an application review is begun. Thus, the process labeled "Collect Information" corresponds to the negotiation phase of Figure 6.3.

A professional evaluation is carried out on each application. This evaluation, labeled "Professional Evaluation" in Figure 6.5, may invoke the subprocess "Refer to Faculty Review" that invites review by an academic department. Depending on the outcome of the faculty review and professional evaluation, one of several other subprocesses may be initiated. The admissions decision may be deferred, awaiting receipt of some additional information. Once information is received, the "Defer Decision" process may result in the applicant being accepted. If an applicant is accepted or rejected, he or she needs to be notified. In the case of acceptance, the "Accept Applicant" subprocess is initiated to follow up with orientation, enrollment, financial aid, and other processes. Finally, an applicant may be referred to a special admissions program.

Figure 6.5 The subprocesses involved in making an admissions decision for each student applicant.

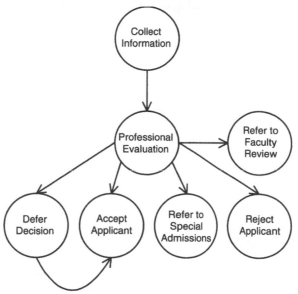

Let's examine our model at this point. Refer to the standard list of questions presented in Chapter 2. Table 6.1 addresses each of these questions in light of the admissions process model so far.

Table 6.1 Standard questions to scrutinize the admissions process model.

What are the relevant subprocesses?	Collect Information Professional Evaluation Refer to Faculty Review Defer Decision Accept Applicant Refer to Special Admissions Reject Applicant
What are the dependencies?	High school transcripts SAT scores Essay High school class rank Final senior grades Faculty interview or audition evaluation
What triggers the process?	Student application received via the Web
Who are the participants?	Student (applicant) Parents or caregivers Guidance counselor High school teachers High school office personnel Admissions office support staff Admissions office professional staff Academic department faculty
What are the inputs?	Complete detailed application submitted by student, including demographic information, educational background, transcripts, SAT scores, and a written essay Application fee received from student Applicant interview notes or evaluation of student audition Teacher and guidance counselor recommendations
What are the outputs?	Current status of application Information that has not yet been received Decision on applicant disposition (reject, accept, refer to special admissions, or defer decision)

Table 6.1 Standard questions to scrutinize the admissions process model. (continued)

What are the business rules, implicit or explicit, that define, support, or limit the process?	Was the application fee paid? Has all requested information been submitted and entered into the system? Has the student requested a specific major that requires academic review of qualifications? Does the student meet the criteria for admission?
What determines that the process is a success?	All received applications are processed based on the inputs received The best candidates are selected for admission The university confirms enrollment of the allowed number of new students
What exceptions or errors may occur?	Student cancels the application Information not received within specified time limits Application fee not received

Note that there are several exceptional conditions that we must take into account. For example, a student may withdraw his or her application. Thus, the application must be marked "Inactive" and any pending reviews cancelled. Also, the application fee or other information, such as transcripts, may not be received in time. The system needs to generate reminders for the "Active" files that certain expected items have not yet been received.

6.5 An Activity Model

To limit the part of the e-admissions system we consider in this case study, consider only the three subprocesses in Figure 6.5 labeled "Collect Information", "Professional Evaluation", and "Refer to Faculty Review". By sitting down with Kelly and reviewing our models, we can outline the steps involved by each of the actors in the Web-based application process. These steps are summarized in the following swimlane activity diagram (Figure 6.6).

Figure 6.6 A swimlane activity diagram illustrating the actors and their activities in processing a student application for admission.

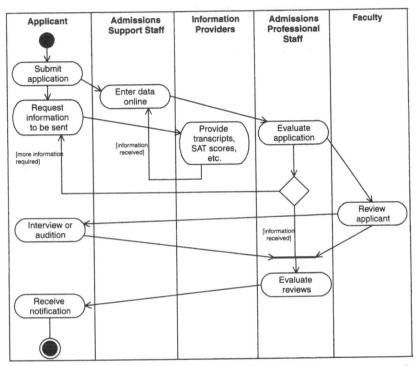

We have divided the actors for this application submission, evaluation, and decision process into five categories. Each actor carries out a number of activities, as detailed in the following text.

First, the applicant completes the online application and submits it to the university Admissions office. The applicant is also responsible for requesting certain information to be forwarded to the Admissions office as part of the application. If requested, the applicant may be called in for an interview, or audition for a faculty review board.

Second, the Admissions office support staff makes sure that each application is complete, the application fee has been received, and data from other sources are entered into the system.

The third actor represents anyone that provides information to the Admissions office. This list may include guidance counselors, teachers, high school staffers, etc.

The fourth participant is the Admissions office professional staff. They are responsible for evaluating applicants SAT scores, essays, GPA, class rank, etc., relative to the university's admission standards. During the application review, the staff may decide that additional information must be requested, or they can initiate a faculty review (interview or audition).

The fifth component is when an academic department faculty member or members optionally review applicants. There is a synchronization point under the "Admissions Professional Staff" swimlane in the diagram. This is the point at which the Admissions professional staff and the optional departmental reviews are complete. This set of information is used to make the final decision on acceptance, and the applicant is notified of the result.

6.6 The State Diagram

Now that we have a good idea of the actors involved, we can concentrate on how we are going to handle an application. A review with Kelly reveals that she wants all of the information about an applicant to appear on one page. This facilitates review by the professional staff, which must look at about one thousand applications each year. To prevent information overload and help organize the data, Kelly agrees that the application can contain expandable/collapsible sections. Fortunately, Domino directly supports the display of document sections, including the ability to collapse or expand them as desired[1].

As we are going to treat each student's application as a single document, we need to model the admissions process from a document-based viewpoint. The natural way to do this in UML is with a state diagram, as we have done in previous chapters. A big advantage in this type of model is the easy transition to the Notes document-based implementation.

Figure 6.7 shows a state diagram that represents the document processing steps. We allow students to create and save a draft application before submitting it to the university. Submission of the online application form initiates the application process. Note that this diagram does not show the exceptional condition we discussed in Table 6.1. Namely, an applicant may choose to withdraw a submitted application from consideration at any time (or state) in the process.

Figure 6.7 A state diagram representing the document states in the admissions process.

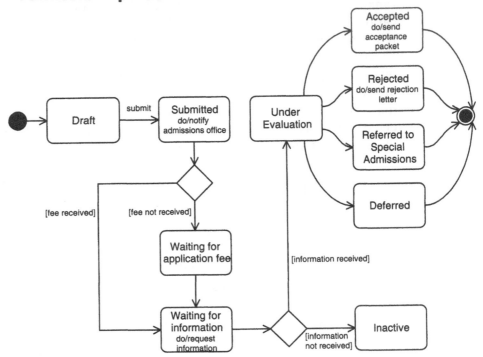

Note that there is a business rule represented by the first decision point in the diagram. This rule prevents the application from proceeding until the application fee has been paid. The next step is to wait for all of the information to be supplied from various providers. This step also includes a business rule to test for the receipt of information. An implicit time limit is associated with this stage of processing. If the time limit is exceeded and requested information is not received, the application is moved to the "Inactive" list. Note that there should be some automated reminders generated to warn the applicant or information provider that a deadline is approaching.

Once the document is under evaluation, it has four target states: "Deferred", "Rejected", "Referred to Special Admissions", and "Accepted". Each of these states is an ending point for this process. However, each of these states may launch other subprocesses to continue handling the application.

6.7 Prototyping the Admissions Database

To create an online application form that will initiate the admissions process, we start by examining the current paper-based form. The basic sections defined in this form are as follows.

- Student demographic information
- Academic major selection
- Personal essay
- Intercollegiate sports
- Activities

These sections must be filled out by the applicant. Note that high school transcripts, GPA, and SAT scores are forwarded by the high school or requested to be sent by the school or testing agency.

6.7.1 Designing the Forms

There are three sections of the application form that ask the applicant to select from lists. These are lists of academic majors, sports activities, and extracurricular activities of interest. Before designing the main form, we first create forms and views so that these are configurable items. They can then be retrieved through a `DbColumn()` formula in the main form. Figures 6.8–6.10 show the form designs. Note that all of these design elements are provided in the `apply.nsf` database on the CD-ROM.

Figure 6.8 The Academic Program form design, showing the programs (majors) categorized by college. The four college categories are defined in a radio button list.

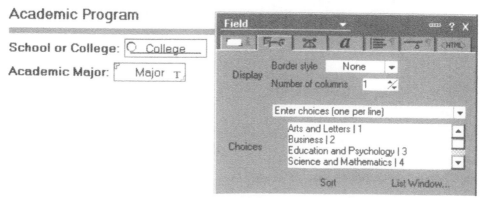

Figure 6.9 The Intercollegiate Sport form design.

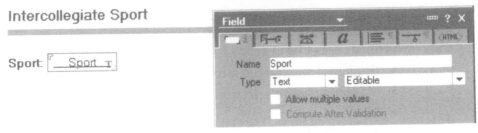

Figure 6.10 The extracurricular Club or Activity form design.

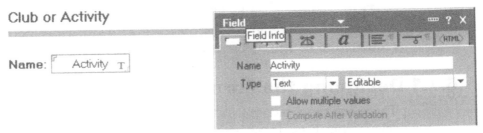

Figures 6.11–6.13 show the views that correspond to each of the three forms, including some example data in each view. The view selection formula for each view simply selects the corresponding documents by form type. Also note that the columns are sorted so they may be retrieved through a lookup.

Figure 6.11 The Academic Programs view design, categorized by college.

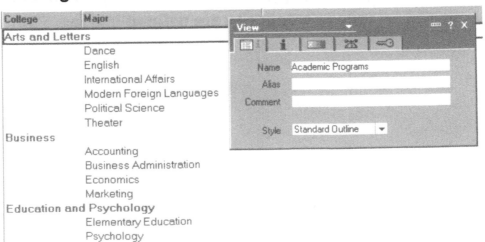

Figure 6.12 The Intercollegiate Sports view design.

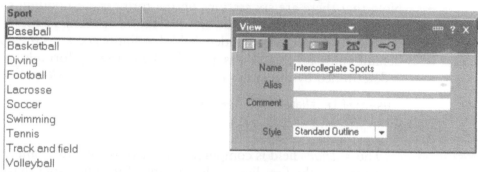

Figure 6.13 The Clubs and Activities view design.

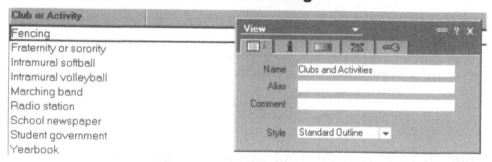

Now we are ready to design the main application form. Remember that our admissions director wants everything for each applicant, including the evaluations by faculty and admissions staff, to be on one page. Thus, we create the following three groups in the Domino directory, and assign the groups to the database access control list, as shown in Table 6.2. We will use these groups to control access to various portions of the form. Thus, only faculty and Admissions office personnel are able to view and edit evaluations. However, anyone can create a draft application and submit it (i.e., the default access is Author).

Table 6.2 Access control groups for the database.

Group	Database ACL
Applicant – person group	Author
Faculty – person group	Editor
Admissions – person group	Editor
Default access	Author

Figure 6.14 shows the design of the Application for Admission form. Note that there are six sections on the form, corresponding to the five sections on the paper form plus one evaluation section. The *AuthorizedReaders* field is a computed Readers field that limits access to the document. The default value for this field is given by the following formula, where the user is determined by the values in the *FirstName* and *LastName* fields.

```
user:=FirstName + " " + LastName;
"Admissions":"Faculty":user
```

The *Authors* field is computed also, with the default value being the user as defined in the first line of the previous formula. Note that both of these fields are hidden, and the form is designated to automatically enable edit mode on open.

To make sure that the evaluation section of the form is not exposed to applicants, this section is created as a controlled access section. The access to the section is designated through the properties window as computed when composed, using the following formula for the groups allowed to view the section.

```
"Admissions":"Faculty"
```

Figure 6.14 The main Application for Admission form.

Wyrcan Flowan University

Application for Admission

Procedures for all applicants:
1. Complete the online application form, filling out all five sections and answering all questions.
2. Submit the application to begin the admissions process.
3. Send a $25 application fee to the Bursar's Office.

▸ **Applicant information**

▸ **Academic major**

▸ **Personal essay**

▸ **Sports**

▸ **Clubs and Activities**

▸ **Applicant Evaluation**

AuthorizedReaders ✏ Authors

Before we address what fields go into each of the six main sections of the form, we need to add some actions to the form. In this initial design, we add three actions. These actions will control saving drafts, submitting the application for evaluation, and saving evaluations. Thus, we need to hide some of these buttons, depending on the document state. Although we haven't introduced it yet, assume there is a field named *Status* in one of the sections.

Table 6.3 shows the action formula and hide-when condition for each action button. Note that the formulas are almost the same, saving the document and displaying a navigator named *DefaultNav*. The only exception is the formula for submitting the application, in which the first line of the formula sets the *Status* field from "Draft" (0) to "Submitted" (1). The hide-when formulas guarantee the following conditions.

- The only action available for a new document is to save a draft.

- An existing draft may either be resaved in draft mode or submitted for evaluation.

- Once submitted, the "Save draft" and "Submit application" actions are no longer available.

- An application under review shows the "Save evaluation" action only.

Table 6.3 The design of the action buttons on the main form.

Action	Formula	Hide-When

Save draft	`@Command([FileSave]);`
	`@Command(`
	`   [OpenNavigator];`
	`   "DefaultNav")`
		`Status != "0"`

Submit application	`FIELD Status:="1";`
	`@Command([FileSave]);`
	`@Command(`
	`   [OpenNavigator];`
	`   "DefaultNav")`
		`@IsNewDoc |`
		`(Status != "0")`

Save evaluation	`@Command([FileSave]);`
	`@Command(`
	`   [OpenNavigator];`
	`   "DefaultNav")`
		`@IsNewDoc |`
		`(Status = "0")`

6.7.2 The Application Subforms

Using the subform capability of Notes is an excellent way to encapsulate changes to each of the six main sections of the application form. Subsequent changes to a subform are automatically reflected in the forms that include the subform. Thus, we do not have to change the design of the Application for Admission form to add or delete fields from any of the input sections of the form, or modify the evaluation section that the Admissions office staff fill out.

Figures 6.15–6.19 are the input sections of the main form. Each figure represents a subform design that is inserted in the main form to create the corresponding named section of the Application for Admission.

Figure 6.15 The *ApplicantInformation* **subform captures demographic information on the applicant. Some of this information, especially the applicant's e-mail address, will be copied to the Domino directory when the applicant is registered as an Internet user.**

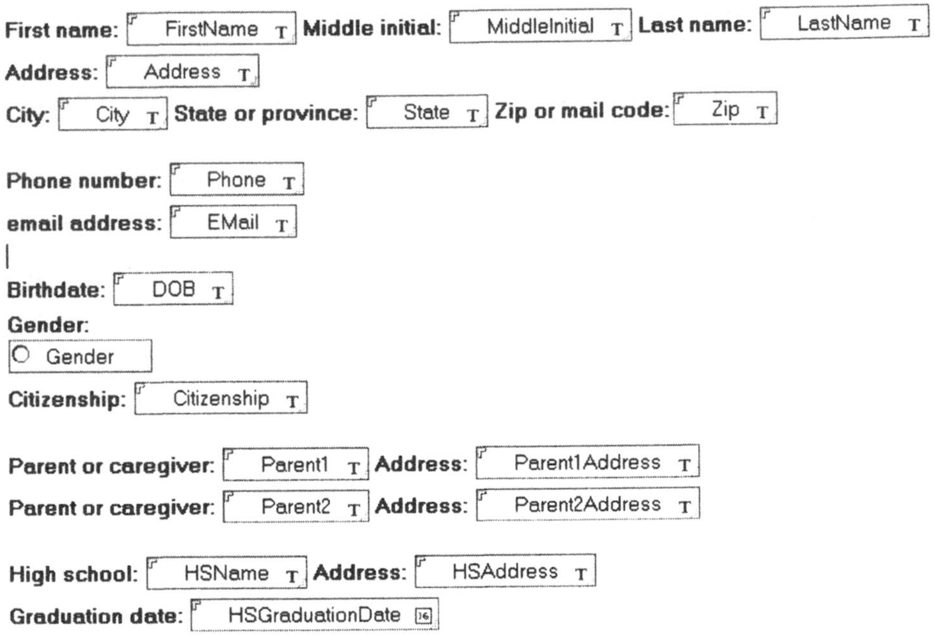

First name: [FirstName T] Middle initial: [MiddleInitial T] Last name: [LastName T]

Address: [Address T]

City: [City T] State or province: [State T] Zip or mail code: [Zip T]

Phone number: [Phone T]

email address: [EMail T]

Birthdate: [DOB T]

Gender:

○ Gender

Citizenship: [Citizenship T]

Parent or caregiver: [Parent1 T] Address: [Parent1Address T]

Parent or caregiver: [Parent2 T] Address: [Parent2Address T]

High school: [HSName T] Address: [HSAddress T]

Graduation date: [HSGraduationDate 📅]

Figure 6.16 The *MajorSelection* **subform allows selection of an academic major, using a radio button with a** `DbColumn` **lookup formula on the "Academic Programs" view.**

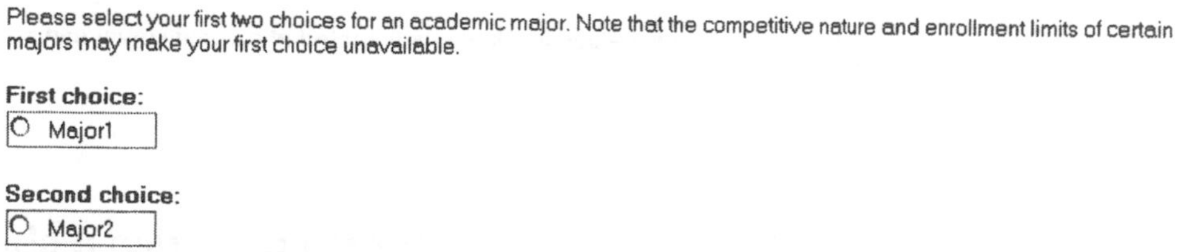
Please select your first two choices for an academic major. Note that the competitive nature and enrollment limits of certain majors may make your first choice unavailable.
First choice:
Major1
Second choice:
Major2

Figure 6.17 The *PersonalEssay* **subform includes a rich text field that allows the applicant to enter an essay. By selecting "Web Access: Display Using Java Applet" in the field properties window, the applicant is provided with a Java applet to allow rich text formatting of his/her essay.**

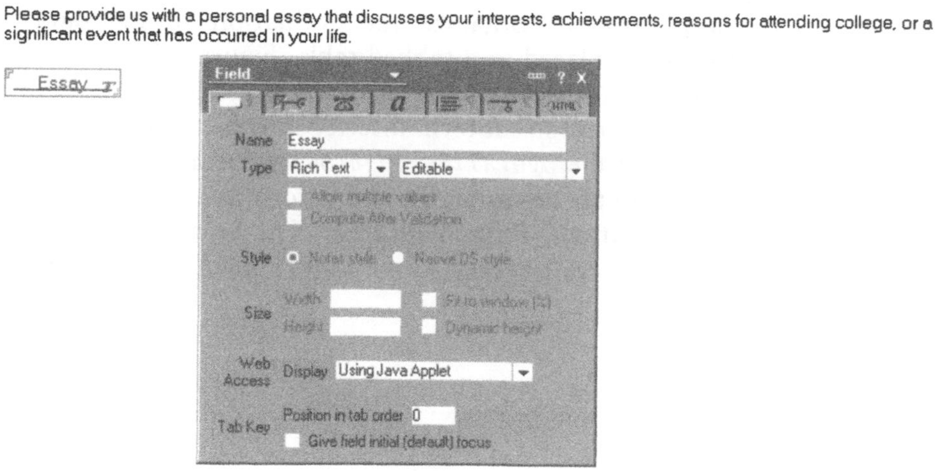
Please provide us with a personal essay that discusses your interests, achievements, reasons for attending college, or a significant event that has occurred in your life.
Essay
Field
Name
Essay
Type
Rich Text
Editable
Style
Size
Web Access
Display
Using Java Applet
Tab Key
Position in tab order
0
Give field initial (default) focus

Figure 6.18 The *SportSelection* **subform presents a check box selection of sports programs, using a** DbColumn **lookup formula on the "Intercollegiate Sports" view.**

Please select any sports for which you have participated at a varsity level and plan to pursue at the University.

☐ VarsitySports

Figure 6.19 The *ActivitySelection* **subform provides a check box selection of clubs and activities through a** DbColumn **lookup on the "Clubs and activities" view.**

Please select any activities that interest you.

☐ Activities

As mentioned in the last section, the evaluation portion of the form is located in a controlled access section so this information is not exposed to the applicant. To organize the evaluation review data, the evaluation subforms are placed in a tabbed table. Figure 6.20 shows the design of the tabbed table inside the controlled access section. This table also shows the *OverallEvaluation* subform inserted into the first tab of the table.

Note that the *OverallEvaluation* subform contains the *Status* field we referred to in the design of the action formulas and hide-when conditions on the action buttons of the main form. The field is designated a radio button, with values corresponding to the states shown in the state diagram of Figure 6.7.

0 Draft
1 Submitted
2 Inactive
3 Deferred
4 Referred to Special Admissions
5 Accepted
6 Rejected

The other radio button fields on the subform are formatted as a horizontal row that allows an Admissions evaluator to score each item on a scale from one to five.

Figure 6.20 The *OverallEvaluation* subform embedded in a tabbed table.

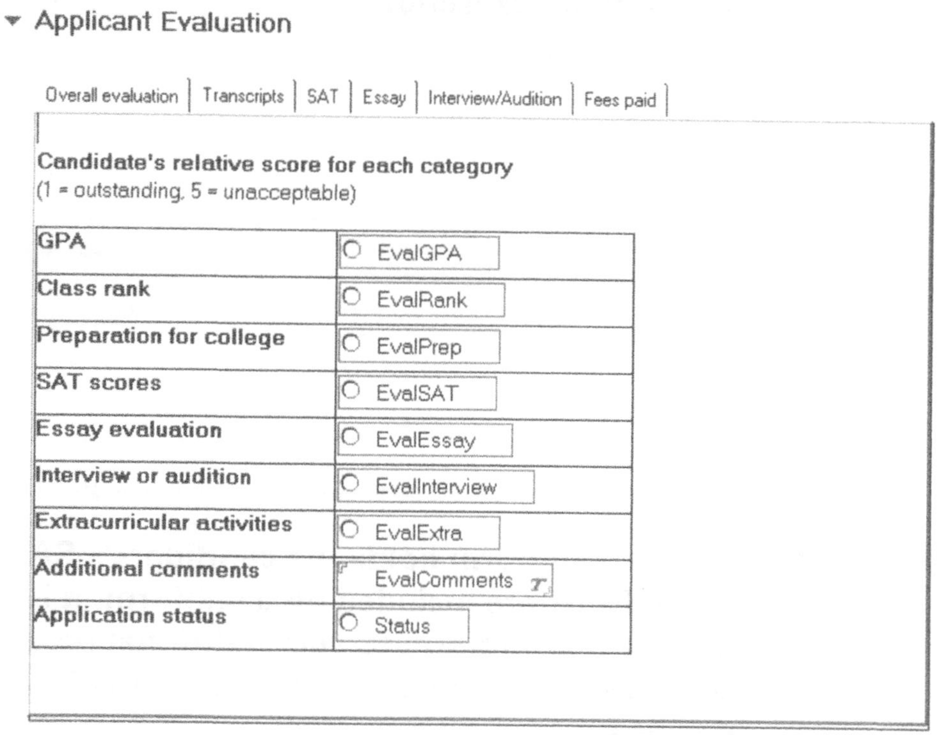

Figures 6.21–6.25 show the remaining evaluation subforms that comprise the other tabs on the table.

Figure 6.21 The *Transcripts* **subform captures the applicant GPA and class rank. It includes a rich text area for containing the high school transcripts.**

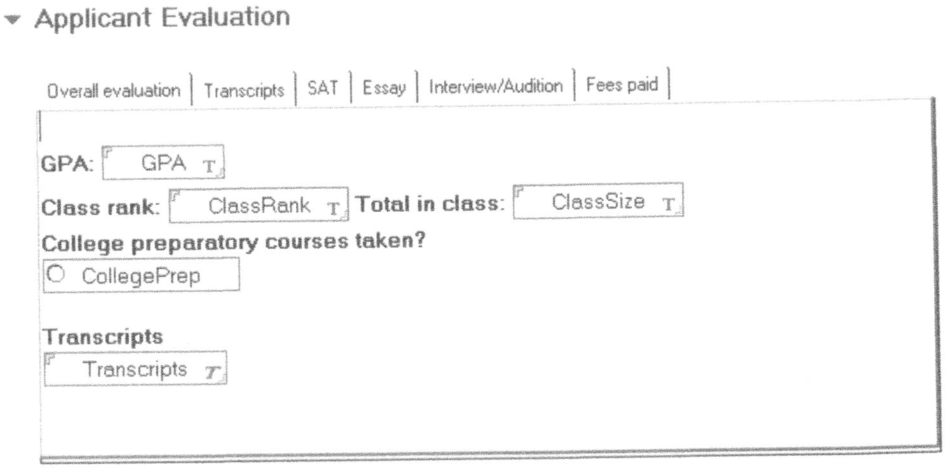

Figure 6.22 The *Tests* **subform holds the candidate's SAT scores. As these fields are included on a subform, it is easy to modify this subform to add additional types of test scores without affecting the rest of the form and view designs.**

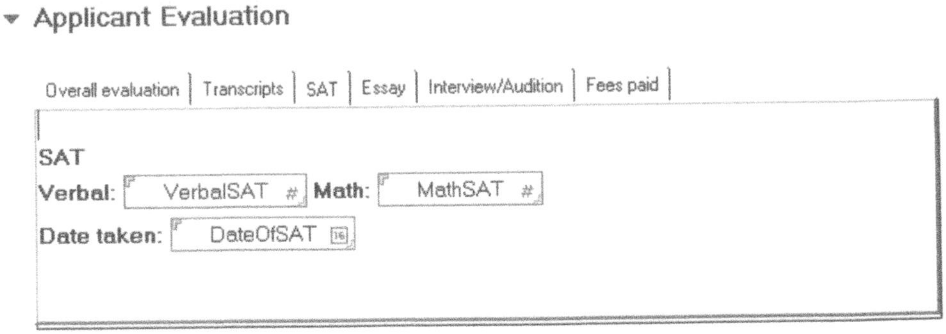

Figure 6.23 To evaluate the student's personal essay, the *EssayEvaluation* **subform includes a rubric to score the paper on several different criteria.**

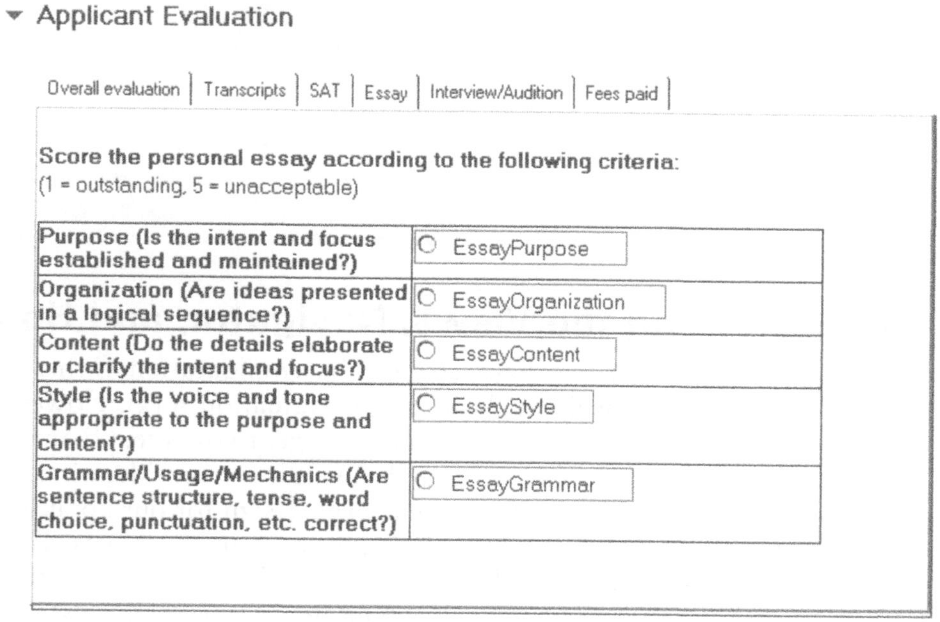

▼ Applicant Evaluation

| Overall evaluation | Transcripts | SAT | Essay | Interview/Audition | Fees paid |

Score the personal essay according to the following criteria:
(1 = outstanding, 5 = unacceptable)

Purpose (Is the intent and focus established and maintained?)	○ EssayPurpose
Organization (Are ideas presented in a logical sequence?)	○ EssayOrganization
Content (Do the details elaborate or clarify the intent and focus?)	○ EssayContent
Style (Is the voice and tone appropriate to the purpose and content?)	○ EssayStyle
Grammar/Usage/Mechanics (Are sentence structure, tense, word choice, punctuation, etc. correct?)	○ EssayGrammar

Figure 6.24 Comments or reviews based on an optional interview or audition are collected using the *InterviewOrAudition* **subform.**

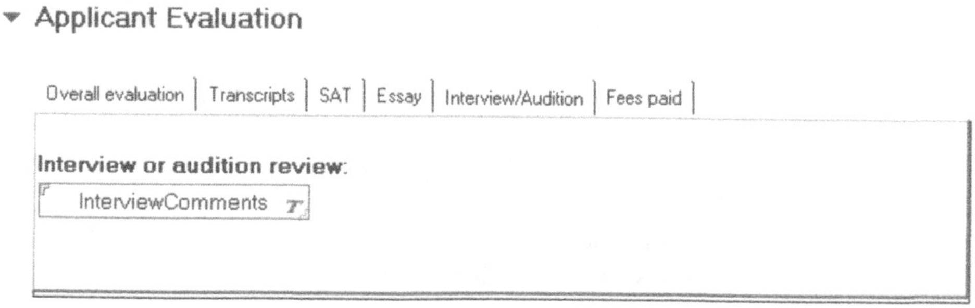

▼ Applicant Evaluation

| Overall evaluation | Transcripts | SAT | Essay | Interview/Audition | Fees paid |

Interview or audition review:

InterviewComments   *T*

Figure 6.25 The *Fees* **subform maintains information on whether application or other types of fees have been paid.**

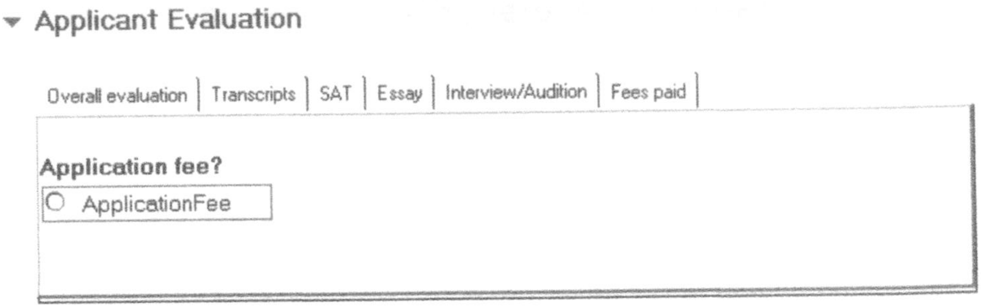

▼ Applicant Evaluation

Overall evaluation | Transcripts | SAT | Essay | Interview/Audition | Fees paid

Application fee?
○ ApplicationFee

6.7.3 Creating the Views, Navigator, and Home Page

At this point, we have created the configuration forms and views, and designed the main application and evaluation form. However, we have not provided a way for a high-schooler to get to the site, fill out a new form, and log in to change their draft and submit it for consideration.

We will show a simple home page design for our application. This home page is created as a Page design element in the Notes database. The two links relevant to our design task are to fill out a new draft application and log in to the site to edit and submit drafts (for applicants) or to log in to the site and evaluate applications (for Admissions staff). Figure 6.26 shows the page, set up as the default page when the database is opened, as it appears in a Web browser.

The hot link for filling out a new draft application is coded in the page as `apply.nsf/Application+for+Admission/OpenForm`. This opens a main application form that can be saved as a new document. The hot link to log in to the site is `apply.nsf/DefaultNav?OpenNavigator&Login`. This opens the same navigator, *DefaultNav*, that we used earlier in the action buttons for the main form.

Figure 6.27 shows the graphical navigator, containing three image-based hot links. The University home page link is designated as a simple "Open URL" action. The "Review applications for admission" link is designated as an "Open a View or Folder" action, opening a view named "Applications".

Finally, the link for editing and submitting a draft is specified to run the following formula.

```
REM "get common name and replace any spaces with plus sign";
commonname:=@Name([CN];@UserName);
user:=@ReplaceSubstring(commonname;" ";"+");
REM "get database name";
dbpath:=@Subset(@DbName;-1);
dbname:=@ReplaceSubstring(dbpath;"\\";"/");
@URLOpen(dbname + "/Draft+Applications/" + user + "?EditDocument&Login")
```

Figure 6.26 The Admissions office home page.

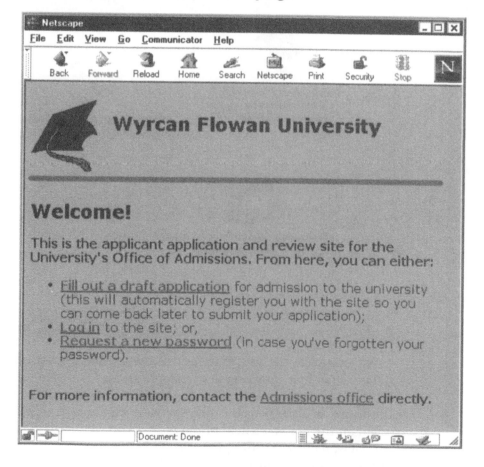

The formula first gets the common name of the user that is logged in to the site. It then retrieves the name of the current database, and uses both of these parameters to construct the next URL to load. This URL retrieves the user's draft document from a view named "Draft Applications", based on the user's name as the key used to select the document.

Figure 6.27 The graphical navigator used by registered users.

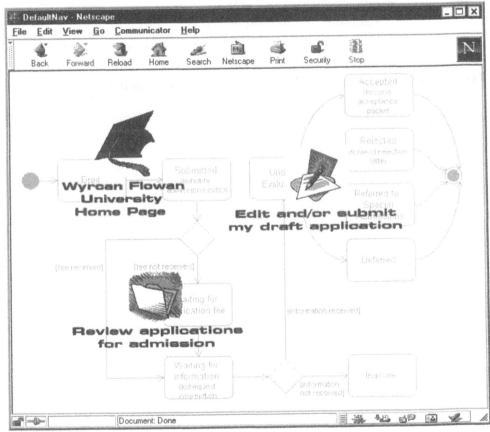

We complete the user interface of the application by designing the two views. The first one, named "Applications", is the one used by the Admissions staff to access each admission application. To organize the applicants in alphabetical order, the view contains an invisible first column based on the *LastName* field of the *ApplicantInformation* subform. The second view, named "Draft applications", is based on the design of the "Applications" view, except that the invisible column for sorting by last name is removed. Instead, the first column contains the common name of the user, to be used as the key in the prevous formula for retrieving an applicant's specific draft document. Figure 6.28 shows a sample view, in which the first visible column is constructed with the formula FirstName + " " + LastName.

Figure 6.28 The "Applications" view as displayed in a Web browser.

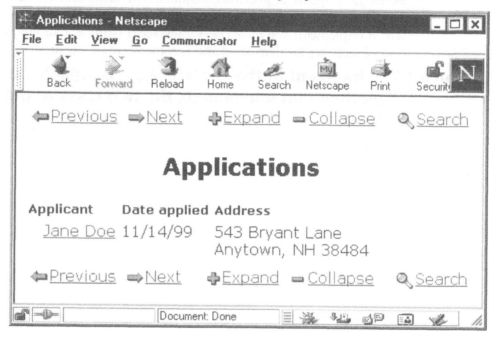

6.7.4 Registering an Applicant

We have one final feature to add to our initial design in this chapter. We need some way to register new applicants in the Domino directory when a new draft application is saved.

Once a new document has been saved as a draft, the system will automatically register the applicant as an Internet user, complete with a system-supplied password. The registration information is then e-mailed to the applicant, who can use the ID to log back in and edit or submit their draft. All of this is accomplished by creating a *WebQuerySave* agent for the Application for Admission form.

To get started, the *WebQuerySave* event of the Application for Admission form is coded with the following formula.

```
@If(@IsNewDoc;@Command([ToolsRunMacro]; "registerApplicant");"")
```

This formula launches a Java agent to carry out the registration tasks for new documents only. The *registerApplicant* agent has the same structure as the Java agents in previous chapters. The tasks carried out by the agents are encapsulated in two classes, *registerTask* and *notifyTask*, each of which implements the *notesTask* interface. Listing 6.1 shows how these two task objects are implemented and run by the main agent.

Listing 6.1 The fragment of the *registerApplicant#NotesMain()*
method that implements the two tasks of the
WebQuerySave **agent.**

```
// carry out the agent tasks - first, register the user.
registerTask register = new registerTask();
register.perform();


// see if the user was newly registered
if ( register.newRegistrationCompleted() ) {
    // get the password and notify the user via email
    String password = register.getPassword();
    notifyTask notify = new notifyTask(password);
    notify.perform();
    notify = null;

}
register = null;
```

The *registerTask* instance is run first to register the new user in the Domino directory and generate an Internet password. If the registration completes successfully, the password is retrieved from the object and passed to the *notifyTask* instance. This second task is responsible for notifying the new user via e-mail of their new user ID and password.

The *perform()* method of *registerTask* is shown in Listing 6.2. This method accesses the document being saved through the document context that is retrieved from the agent context. This method reuses the *dominoDirectoryLookup* class we developed in Chapter 4 to make sure a duplicate Person entry is not created in the Domino directory. An instance of a new class, *dominoDirectory* (Listing 6.3), is used to actually create the Person document and store in it some of the information captured from the fields on the *ApplicantInformation* subform. The last part of the method saves the new Person document, gets the password generated by the *dominoDirectory#savePerson()* method, and adds the new user to the Applicants group.

Listing 6.2 The *registerTask#perform()* method.

```
public void perform()
{

   try {

      // retrieve the document from the agent context and create a
      // Domino directory interface object
      Document doc = registerApplicant.getContext().getDocumentContext();
      dominoDirectory dir = new dominoDirectory(
         registerApplicant.getAgentSession() );


      // get the user's name
      String firstname = doc.getItemValueString("FirstName");
      String initial = doc.getItemValueString("MiddleInitial");
      String lastname = doc.getItemValueString("LastName");


      // get the email address
      String loginID = firstname + " " + lastname;
      String email = doc.getItemValueString("EMail");


      // check for a duplicate entry
      dominoDirectoryLookup dirlookup = new dominoDirectoryLookup(
         registerApplicant.getAgentSession() );
      if (dirlookup.lookup(loginID) == dominoDirectoryLookup.PERSON)
         return;
```

Listing 6.2 The *registerTask#perform()* **method. (continued)**

```
    // create a new Person document and fill out the appropriate fields
    if ( dir.createPerson(firstname, initial, lastname) ) {
        dir.setPersonInformation( dominoDirectory.PHONE,
                                  doc.getItemValueString("Phone") );
        dir.setPersonInformation( dominoDirectory.EMAIL, email);
        dir.setPersonInformation( dominoDirectory.ADDRESS,
                                  doc.getItemValueString("Address") );
        dir.setPersonInformation( dominoDirectory.CITY,
                                  doc.getItemValueString("City") );
        dir.setPersonInformation( dominoDirectory.STATE,
                                  doc.getItemValueString("State") );
        dir.setPersonInformation( dominoDirectory.ZIP,
                                  doc.getItemValueString("Zip") );
        dir.setPersonInformation( dominoDirectory.COUNTRY,
                                  doc.getItemValueString("Citizenship") );


        // save the document and get the password assigned
        myPassword = dir.savePerson();


        // add the user to the Applicants group
        dir.addToGroup(loginID, "Applicants");
        success = true;
    }


    // clean up
    dir = null;
  }
catch (Exception e) {
    notesLogFile.getInstance().write( "registerTask.perform(): " +
                                      e.toString() );


  }
}
```

The methods in *dominoDirectory* allow selection from among multiple Domino directory databases. Once a specific database is selected, methods are called to create new documents based on the Person form, set specific fields in a new Person document, generate an Internet password for the user, and add a user to a specific group.

Listing 6.3 The *dominoDirectory* class.

```java
// *******************************************************************************
//
// Module: dominoDirectory.java
// Author: Dick Lam
//
// Description: This is a Java class used for interfacing to the Domino
//              directory.
//
// *******************************************************************************


import java.util.*;
import lotus.domino.*;


// *******************************************************************************


public class dominoDirectory
{
    // constructor
    public dominoDirectory(Session s)
    {
        try {
            // get the Domino directories known to this session and set the
            // current database to the first one in the list
            myNABs = s.getAddressBooks();
            if ( (myNABs == null) || myNABs.isEmpty() )
                curDB = null;
            else
                curDB = (Database)myNABs.elementAt(0);
        }
```

Listing 6.3 The *dominoDirectory* **class. (continued)**

```
      catch (Exception e) {
         notesLogFile.getInstance().write( "dominoDirectory: " +
                                           e.toString() );

      }
   }


   // ------------------------------------------------------------------


   // getAddressBookList - returns the list of address books
   public Vector getAddressBookList()
   {
      return myNABs;
   }


   // ------------------------------------------------------------------


   // setCurrentDirectory - sets the current Domino directory
   public void setCurrentDirectory(int index)
   {
      if ( (myNABs != null) && (index >=0) && ( index < myNABs.size() ) )
         curDB = (Database)myNABs.elementAt(index);
   }


   // ------------------------------------------------------------------


   // createPerson - creates a new Person document in the current
   //                directory (returns true if the document was created
   //                okay, false otherwise)
```

Listing 6.3 The *dominoDirectory* **class. (continued)**

```
public boolean createPerson(String first, String initial, String last)
{
    try {
        // make sure the current directory is valid
        if (curDB == null)
            return false;

        // make sure the database is open
        if ( !curDB.isOpen() )
            curDB.open();

        // create a new document based on the Person form
        curPerson = null;
        curPerson = curDB.createDocument();
        curPerson.replaceItemValue("Form", "Person");
        curPerson.replaceItemValue("Type", "Person");

        // save the name
        curPerson.replaceItemValue(FIRSTNAME, first);
        curPerson.replaceItemValue(MIDDLEINITIAL, initial);
        curPerson.replaceItemValue(LASTNAME, last);

        // save the full and short names
        String name = first + " " + last;
        curPerson.replaceItemValue("FullName", name);
        curPerson.replaceItemValue("ShortName", name);

        // set other fields and add the name to the Applicants group
        curPerson.replaceItemValue("MailSystem", "5");
        return true;
    }
```

Listing 6.3 The *dominoDirectory* **class. (continued)**

```
        catch (Exception e) {
          notesLogFile.getInstance().write( "dominoDirectory#createPerson: " +
                                              e.toString() );

          return false;
        }
    }



    // -----------------------------------------------------------------



    // setPersonInformation - sets specific information into the current
    //                        person document
    public void setPersonInformation(String field, String value)
    {
      try {
        curPerson.replaceItemValue(field, value);
      }
      catch (Exception e) {
        notesLogFile.getInstance().write(
          "dominoDirectory#setPersonInformation: " + e.toString() );
      }
    }



    // -----------------------------------------------------------------



    // savePerson - saves the current person document (returns the Internet
    //              password generated for the person)
    public String savePerson()
    {
      try {
        // generate a password and set it in the password field
        String password = generatePassword();
        curPerson.replaceItemValue("HTTPPassword", password);
```

Listing 6.3 The *dominoDirectory* **class. (continued)**

```
      // save the document and return
      if ( !curPerson.computeWithForm(true, true) )
         throw new Exception("document validation failed");
      if ( !curPerson.save(true) )
         throw new Exception("document was not saved");


      return password;
   }
   catch (Exception e) {
      notesLogFile.getInstance().write( "dominoDirectory#savePerson: " +
                                        e.toString() );
      return null;
   }
}


// --------------------------------------------------------------------


// addToGroup - adds a new member name to a group (returns true if the
//              name is already in the list, or if the name was added,
//              false otherwise)
public boolean addToGroup(String memberName, String groupName)
{
   try {
      // make sure the current directory is valid
      if (curDB == null)
         return false;

      // make sure the database is open
      if ( !curDB.isOpen() )
         curDB.open();
```

Listing 6.3 The *dominoDirectory* **class. (continued)**

```
        // open the named group document
        View groups = curDB.getView("Groups");
        Document groupdoc = groups.getDocumentByKey(groupName);


        // get the member list and add the new member (test to make sure
        // the member is not already in the list)
        Vector members = groupdoc.getItemValue("Members");
        for (int i = 0; i < members.size(); i++) {
            String test = (String)members.elementAt(i);
            if ( test.equalsIgnoreCase(memberName) )
                return true;
        }


        // the name is not already in the list - add it
        members.addElement(memberName);
        groupdoc.replaceItemValue("Members", members);


        // save the updated group document
        groupdoc.save(true);
        return true;
    }
    catch (Exception e) {
        notesLogFile.getInstance().write(
            "dominoDirectory#addToGroup: " + e.toString() );
        return false;
    }
}


// -------------------------------------------------------------------
```

Listing 6.3 The *dominoDirectory* **class. (continued)**

```
// generatePassword - generates a new password
private String generatePassword()
{
   try {
      // base the password on the document id
      String unid = curPerson.getUniversalID();
      StringBuffer buf = new StringBuffer();
      for (int i = 0; i < 32; i += 4) {
         buf.append( unid.charAt(i) );
      }

      return new String(buf);
   }
   catch (Exception e) {
      notesLogFile.getInstance().write(
         "dominoDirectory#generatePassword: " + e.toString() );
      return new String("");
   }
}



// **************************************************************************
// properties

private Vector myNABs = null;        // list of Domino directory databases
private Database curDB = null;        // current database

private Document curPerson = null;   // current Person document

// field names
public static final String FIRSTNAME = "FirstName";
public static final String MIDDLEINITIAL = "MiddleInitial";
public static final String LASTNAME = "LastName";
public static final String PHONE = "PhoneNumber";
```

Listing 6.3 The *dominoDirectory* class. (continued)

```
    public static final String EMAIL = "MailAddress";
    public static final String ADDRESS = "StreetAddress";
    public static final String CITY = "City";
    public static final String STATE = "State";
    public static final String ZIP = "Zip";
    public static final String COUNTRY = "Country";

}


// ***********************************************************************


// end of dominoDirectory.java
```

To notify the user that he/she is a registered user with a user Id and password, the *notifyTask#perform( )* method (Listing 6.4) is called by the main agent. This method reuses the notification database from Chapter 4 to handle the e-mail notification, based on the e-mail address provided by the user on the Application for Admission form.

Listing 6.4 The *notifyTask* class.

```
// ***********************************************************************
//
// Module: notifyTask.java
// Author: Dick Lam
//
// Description: This is a Java class that implements the notesTask interface.
//              The task of this class is to notify a user of his/her
//              registration user id and password.
//
// ***********************************************************************


import java.util.*;
import lotus.domino.*;
```

Listing 6.4 The *notifyTask* class. (continued)

```
// ***********************************************************************

public class notifyTask implements notesTask
{
    // constructor
    public notifyTask(String password)
    {
        myPassword = new String(password);
    }


    // -------------------------------------------------------------------


    // perform - performs this task
    public void perform()
    {
        try {
            // retrieve the document from the agent context
            Document doc = registerApplicant.getContext().getDocumentContext();

            // get the recipient's name and set up the text of an email
            // message that will be sent to the user
            String loginID = doc.getItemValueString("FirstName") + " " +
                            doc.getItemValueString("LastName");
            StringBuffer buf = new StringBuffer();
            buf.append("Thank you for registering with the Wyrcan Flowan\n");
            buf.append("University Admissions Office. When you registered\n");
            buf.append("by saving your draft application for admission, we\n");
            buf.append("saved your draft and generated the following user\n");
            buf.append("ID and password so you may re-enter the site and\n");
            buf.append("continue editing your application. Your user ID\n");
            buf.append("and password are:\n\n");
            buf.append("    User ID: ").append(loginID).append("\n");
            buf.append("    password: ").append(myPassword).append("\n\n");
```

Listing 6.4 The *notifyTask* class. (continued)

```
buf.append("Note that the Admissions office will not be able to\n");
buf.append("review your application until you submit it for\n");
buf.append("consideration, and pay the application fee.\n");
String text = new String(buf);

// create a notification entry
Database notifyDB = findDB();
if (notifyDB == null)
    throw new Exception("could not find notification database");

notifyDB.open();
Document notifyDoc = notifyDB.createDocument();
notifyDoc.replaceItemValue("Form", "Notification Entry");
notifyDoc.replaceItemValue("RequestType", "1");
notifyDoc.replaceItemValue("Requestor", "WFU Admissions Office");
notifyDoc.replaceItemValue("Actors", loginID);
notifyDoc.replaceItemValue("Subject", "New user information");
notifyDoc.replaceItemValue("NotificationFrequency", "1");
notifyDoc.replaceItemValue("EntryBody", text);

// save the document
if ( !notifyDoc.computeWithForm(true, true) )
    throw new Exception("notification document validation failed");
if ( !notifyDoc.save(true) )
    throw new Exception("notification document was not saved");

// clean up and return
notifyDoc = null;
notifyDB = null;
}
```

Listing 6.4 The *notifyTask* **class. (continued)**

```
        catch (Exception e) {
            notesLogFile.getInstance().write( "notifyTask#perform(): " +
                                          e.toString() );
        }
    }


    // -------------------------------------------------------------------

    // findDB - finds the notification database
    private Database findDB()
    {
        try {
            // look through the server database directory for the
            // notification database
            DbDirectory dir =
                registerApplicant.getAgentSession().getDbDirectory(null);
            Database db = dir.getFirstDatabase(DbDirectory.DATABASE);
            while (db != null) {
                // check the title of the database
                if ( db.getTitle().equalsIgnoreCase("Workflow Notifier") )
                    return db;

                // not found - try the next one
                db = dir.getNextDatabase();
            }

            return null;
        }
        catch (Exception e) {
            notesLogFile.getInstance().write( "notifyTask#findDB(): " +
                                          e.toString() );
            return null;
        }
```

Listing 6.4 The *notifyTask* **class. (continued)**

```
    }

    // *****************************************************************
    // properties

    private String myPassword = null;
}

    // *****************************************************************

// end of notifyTask.java
```

The *findDB()* method is used to find the "Workflow Notifier" database on the server, using the *lotus.domino.DbDirectory* class. Once the database is found, a new Notification Entry document is created to hold the e-mail notification message that will be sent to the new user. Once the document is saved, the agents in the notification database take care of sending out the e-mail message. Notice how this encapsulation of functionality greatly simplifies the task of the *WebQuerySave* agent.

6.8 Summary

We introduced a case study of a college admissions process. An example admissions office process for handling student applications was introduced and modeled using the workflow modeling approach introduced in Chapter 2. The activity and state diagrams for a portion of the admissions procedures were given, and a sample database was prototyped. This provides a starting point for a Web-based workflow implementation for the processing of student admission applications.

The next chapter will extend the design of the admissions database by adding the components of our workflow framework.

6.9 References

1   F. Collins, et.al. January, 1999. *Lotus Domino Release 5.0: A Developer's Handbook*. IBM International Technical Support Organization.

Chapter 7

Implementing the
Workflow Model

We've arrived at our last chapter. If you've followed us through from Chapter 1, you have seen the basics of process engineering and workflow. We've seen how to create blueprints for a workflow process through UML modeling, and reviewed the Notes/Domino architecture from a workflow perspective. We built a basic notification engine and a generic framework for workflow-enabling an application. We presented a case study of a college admissions process, along with a basic supporting application. Now we'll tie all of the preceding sections together by implementing our workflow framework in the college admissions application developed in Chapter 6.

In this chapter, we cover the changes necessary to implement the workflow framework in `apply.nsf`. We review the configuration documents that make up our workflow engine, and discuss the agents that provide the processing power. We'll finish up by taking the workflow-enabled college application form for a short test drive through the first few steps in its workflow process.

7.1 First Steps

To begin, let's walk through the steps of moving our workflow management system (WFMS) framework design elements from `chap5.nsf` into the admissions database (`apply.nsf`) of Chapter 6. These steps have already been completed, and you can see the end result in the `chapter7.nsf` database (available on the CD-ROM). However, the checklist in Table 7.1 will be valuable when integrating the framework into your own applications.

Table 7.1 Framework implementation tasks.

Copy
Forms
`wf_Action`
`wf_Profile`
`wf_Role`
`wf_State`
Subforms
`wf_Config`
`wf_Subform`
Views
`Workflow\Actions`
`Workflow\wfActionsLookupByName`
`Workflow\wfActionsLookupByNumber`
`Workflow\Profiles`
`Workflow\Roles`
`Workflow\wfRolesLookupByName`
`Workflow\wfRolesLookupByNumber`
`Workflow\States`
`Workflow\wfStatesLookupByName`
`Workflow\wfStatesLookupByNumber`

Table 7.1 Framework implementation tasks. (continued)

Copy

Agents
wfGetCurActionsList
wfGetFormList
wfProcess
wfProcessForm
wfOverDueMonitor
Script Library
wfScriptLibrary

The design elements in Table 7.1 represent all of the necessary components that integrate our WFMS framework with any Domino application. You can create your own interface for the creation and maintenance of the framework documents, or you may choose to use the frameset and outline that we created in the chap5.nsf (Figure 7.1). For the Wyrcan Flowan University admissions application, we have chosen to reuse this interface, but it is an optional part of the WFMS framework.

Figure 7.1 The WFMS framework interface.

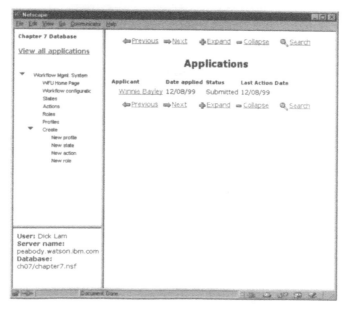

We have attempted to leave the best features of the WFU admissions application developed in the last chapter as we integrated the workflow engine. However, as with any application, some changes were inevitably necessary. We will start describing the changes made to apply.nsf, and then walk through the creation of all documents that our workflow engine needs to implement the design found in Figure 6.7.

Starting with the WFU home page, we had to modify the WFU navigator (DefaultNav). In apply.nsf, the "Review applications for admission" link took us to the "Applications" view. In chapter7.nsf, that link has been changed to the following hotspot formula.

```
@URLOpen("wfFrameset?OpenFrameset&Login")
```

This link lets us access the wfAdmin frameset for creating, viewing, and maintaining the workflow configuration components.

We also made several changes to the wfAdmin outline. An outline entry entitled "WFU Home Page", along with a formula that opens up the original Wyrcan Flowan University home, has been added. Also, a new link appears at the top of the outline page wfOutlinePage. "View all applications" displays the "Applications" view, which is the default view for this frameset.

Figure 7.2 The Wyrcan Flowan University default navigator.

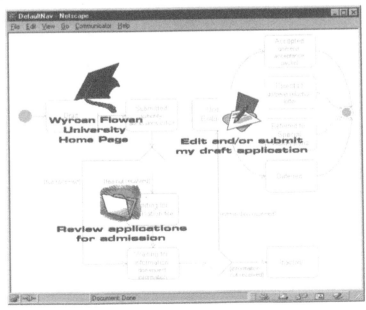

The basic `apply.nsf` database used a *WebQuerySave* agent named *registerApplicant* to register an applicant as a user on the site. This registration was done during each applicant's initial edit session on an application form. Subsequent processing on the document was handled by manually setting the status of the application, ignoring the *registerApplicant* agent.

In our new version, this functionality must change so we can still utilize the *registerApplicant* agent for each applicant's first editing session on an application. But, we must also provide for the execution of the workflow engine agents. To accomplish this, we create a new field on the workflow subform (`wfSubForm`) called *wfAgentToRun*. This field has the following formula.

```
@If(@IsNewDoc;"(registerApplicant)";"(wfProcess)")
```

This formula returns the appropriate agent name, depending on whether a document is new or not. The agent name is made available to the *WebQuerySave* event. This event uses the agent name in the following formula.

```
@Command([ToolsRunMacro]; wfAgentToRun)
```

The syntax of the *WebQuerySave* event is somewhat limiting, so this drove the need for an additional field on the subform. Also, to make use of the *registerApplicant* agent, we need to override the workflow engine for the initial session. This requires the addition of a new State, "New Draft", which is the default state for all new applications. When a user completes their initial editing session on their application, the only option available to them is a "Submit" button. This button is the default "Submit" button on a Domino form, so it is independent of the *wfActions* radio button options displayed by the workflow subform.

Note that in Chapter 5, we did not hide the subform — we wanted it to be visible to illustrate its use in the example. In the `chapter7.nsf` database, the subform is hidden by a Hide paragraph formula of `@IsNewDoc`. When the user clicks "Submit" during their first session, the *registerApplicant* agent executes and saves the document. When the user accesses the document for a second time by logging into the site with their user ID and password, the workflow subform is now visible since it's not a new document. And, as the user has logged in for this second session (or subsequent sessions), *wfSubform* can use the authenticated user information. Of course, in a real rollout of a similar application, you can choose to hide the subform all of the time if desired.

The *wfState_num* field on the workflow subform is set to have a default value of 2, corresponding to the "Draft". To reflect the changes to the registration process, we have included some additional introductory text at the top of the "Application for Admission" form to explain the process to the user. This text is only visible during the first editing session.

Note that all of the process steps are triggered either by the initial "Submit" button or the selection of an action through the radio button action choices presented by the workflow subform. Thus, we deleted all of the actions ("Submit Application", "Save Draft", "Save Evaluation") that appeared in the action bar of the original form.

The next change is in the formula for the "Status" field that appears on the "Overall Evaluation" tab of the applicant evaluation section. This was changed from a series of radio buttons (presenting the choices "Draft", "Submitted", "Inactive", "Deferred", "Referred to Special Admissions", "Accepted", and "Rejected") to the following formula.

```
@If(@IsNewDoc;"New Draft"; @DbLookup("":"NoCache";"";
    "Workflow\\wfStatesLookupByNumber";wfState_num;2))
```

This field will now display the state name corresponding to the state number in the workflow subform, or simply "New Draft" if this is a new document.

Finally, in the applicant evaluation section, we added a field to the "Transcripts" tab called "InfoComplete". This field will let the support staff indicate when all supporting information and documents have been received. This field, along with the radio button field for determining if the application fee is paid, will be used by our XML rules in the "Proceed" action to determine the next appropriate state for the document as it progresses through the workflow.

7.2 Configuring the Workflow

Now let's walk through the workflow engine configuration documents. We'll start with the profile document. From our Wyrcan Flowan University home page, click on "Workflow Administration" to change to the workflow interface. Click on the "Profiles" link. The view is displayed in the main frame, consisting of just one document. Opening this document will reveal the following settings.

Table 7.2 Profile document settings.

Workflow name	WFUAdmissions
Workflow administrator access group	Administrators
History log server	Local
History log DB file	wfLog.nsf
Roles/Groups NAB server	Local
Roles/Groups NAB database	Names.nsf
Notification DB server	Local
Notification DB	Notify.nsf

We have: created a name for the workflow process of "WFUAdmissions"; identified a Notes ACL group that can serve as overall administrator for the workflow; and specified the names and locations of the history log, Domino directory, and Notification database.

7.2.1 The State Documents

With the exception of adding the "New Draft" state mentioned in the last section, the states for the workflow remain the same as in the Chapter 6 design (Table 7.3).

Table 7.3 The states of the workflow process.

New Draft	Accepted
Draft	Rejected
Submitted	Referred to Special Admissions
Waiting for Application Fee	Deferred
Waiting for Information	Inactive
Under Evaluation	

The state document for the "New Draft" state (Table 7.4) is more for documentation purposes than functionality, as the workflow engine starts with a default state of "Draft". The "New Draft" state is used only for the first editing session, before the workflow engine starts.

Table 7.4 The "New Draft" state.

Workflow name	WFUAdmissions
Form name	Application for Admission
State name	New Draft
State number	1
State owner	Applicant
State duration	1
Send overdue notification?	No
State overdue notice text	
Send overdue follow-up?	No
Overdue follow-up duration	
Overdue follow-up text	

Table 7.5 shows the "Draft" state document. The owner for this state is the applicant. We have set a one-month duration for this state, and developed the text of the notification e-mail if that duration is exceeded. Given the nature of the document and state, we do not set the overdue follow-up switch.

Table 7.5 The "Draft" state.

Workflow name	WFUAdmissions
Form name	Application for Admission
State name	Draft
State number	2
State owner	Applicant
State duration	1 Month
Send overdue notification?	Yes

Table 7.5 The "Draft" state. (continued)

State overdue notice text	Dear WFU Applicant, Recently you started the online application process for admission to WF University. We thank you for your interest in WFU and look forward to processing your application. In order to process your application, you need to log into the WFU website again, using the user name and password that were e-mailed to you, and submit the application. Applications that are not submitted 30 days after being created online are automatically deleted. Again, thank you for your interest in WFU. Admissions Office
Send overdue follow-up?	No
Overdue follow-up duration	
Overdue follow-up text	

The "Submitted" state (Table 7.6) specifies the admissions office support staff as the owner. It is that group's responsibility to coordinate the reception of admissions material and record payment of the application fee. Only then is the application passed on for evaluation by the professional staff. A short notification and follow-up message is specified to send to the admissions office employees.

Table 7.6 The "Submitted" state.

Workflow name	WFUAdmissions
Form name	Application for Admission
State name	Submitted
State number	3
State owner	Admissions Support Staff
State duration	1 Month
Send overdue notification?	Yes
State overdue notice text	This document has exceeded the 30-day limit for Submitted documents. Please process the document immediately.
Send overdue follow-up?	Yes

Table 7.6 The "Submitted" state. (continued)

Overdue follow-up duration	1 Week
Overdue follow-up text	********ATTENTION********* This document has exceeded the 30-day limit for Submitted documents. Please process the document immediately.

The "Accepted" state (Table 7.7) is one of the termination points for the scope of our workflow process. Normally, this would be the starting point for a number of other subprocesses, such as acceptance package mailings, housing preferences, financial aid applications, etc.

As processing will stop at this point, we leave the ownership with the admissions professional staff and forgo any notification messages.

Table 7.7 The "Accepted" state.

Workflow name	WFUAdmissions
Form name	Application for Admission
State name	Accepted
State number	4
State owner	Admissions Professional Staff
State duration	
Send overdue notification?	No
State overdue notice text	
Send overdue follow-up?	No
Overdue follow-up duration	
Overdue follow-up text	

Another termination point is the "Rejected" state (Table 7.8). No duration or follow-up are necessary, and the ownership remains with the professional staff.

Table 7.8 The "Rejected" state.

Workflow name	WFUAdmissions
Form name	Application for Admission
State name	Rejected
State number	5
State owner	Admissions Professional Staff
State duration	

Table 7.8 The "Rejected" state. (continued)

Send overdue notification?	No
State overdue notice text	
Send overdue follow-up?	No
Overdue follow-up duration	
Overdue follow-up text	

The "Waiting for Application Fee" state (Table 7.9) is entered when the application is submitted and the transcripts, etc., have been received, but the application fee payment has not been made. In this case, ownership is transferred back to the applicant so he/she can receive notice of the tardy funds.

Table 7.9 The "Waiting for Application Fee" state.

Workflow name	WFUAdmissions
Form name	Application for Admission
State name	Waiting for Application Fee
State number	6
State owner	Applicant
State duration	1 Month
Send overdue notification?	Yes

State overdue notice text	Dear WFU Applicant,
	    As of today, we have not received the $25 Application fee that is necessary to process your application. Please forward a check or money order to the WFU Admissions Office. If you have recently sent your payment, please disregard this letter.
	    WFU Admissions Office

Send overdue follow-up?	Yes

Table 7.9 The "Waiting for Application Fee" state. (continued)

Overdue follow-up duration	1 Month

Overdue follow-up text	******Second Notice************
	Dear WFU Applicant,
	    As of today, we have not received the $25 Application fee that is necessary to process your application. Please forward a check or money order to the WFU Admissions Office. If we do not receive your payment within 10 days, your application for admission to WFU will be deleted. If you have recently sent your payment, please disregard this letter.
	    WFU Admissions Office

The document moves to the "Waiting for Information" state (Table 7.10) if the application is submitted but the supporting materials (transcripts, SAT scores, etc.) have not been received. In this case, state ownership is again transferred back to the applicant for notification of the missing documentation.

Table 7.10 The "Waiting for Information" state.

Workflow name	WFUAdmissions
Form name	Application for Admission
State name	Waiting for Information
State number	7
State owner	Applicant
State duration	1 Month
Send overdue notification?	Yes

State overdue notice text	Dear WFU Applicant,
	    As of today, we have not received the supporting materials (transcripts, SAT scores, etc.) that are necessary to process your application. Please contact your high school guidance office or guidance counselor to inform them of this situation.
	    WFU Admissions Office

Send overdue follow-up?	Yes

Table 7.10 The "Waiting for Information" state. (continued)

Overdue follow-up duration	1 Month

Overdue follow-up text	******Second Notice************
	Dear WFU Applicant,
	    As of today, we have not received the supporting materials (transcripts, SAT scores, etc.) that are necessary to process your application.
	    Please contact your high school guidance office or guidance counselor immediately to inform them of this situation.
	    WFU Admissions Office

The state named "Under Evaluation" (Table 7.11) is the heart of the admissions workflow process. The application is completed, the application fee has been received, and the supporting information has arrived. Ownership of the application moves to the admissions professional staff. We indicate a three-month time period as a duration; a sign that the office anticipates heavy volumes and a potential backlog of applications. We include an overdue notification and a follow-up to make sure that no applications slip through the cracks.

Table 7.11 The "Under Evaluation" state.

Workflow name	WFUAdmissions
Form name	Application for Admission
State name	Under Evaluation
State number	8
State owner	Admissions Professional Staff
State duration	3 Months
Send overdue notification?	Yes

State overdue notice text	An Application has exceeded its time period for being "Under Evaluation".
	    Please access the Applications database and process this document.
	    Thank you.

Send overdue follow-up?	Yes

Table 7.11 The "Under Evaluation" state. (continued)

Overdue follow-up duration	1 Month
Overdue follow-up text	This is a follow-up to a previously sent overdue notice. This application continues to be overdue for processing. Please process it immediately. Thank you.

The "Deferred" state (Table 7.12) gives the admissions professional staff the ability to keep a borderline application active until some other situation occurs (e.g., a lower than expected number of quality applications, the student's completion of a summer course, etc.). This state has a lengthy duration of twelve months, with no notification or follow-up.

Table 7.12 The "Deferred" state.

Workflow name	WFUAdmissions
Form name	Application for Admission
State name	Deferred
State number	9
State owner	Admissions Professional Staff
State duration	12 Months
Send overdue notification?	No
State overdue notice text	
Send overdue follow-up?	No
Overdue follow-up duration	
Overdue follow-up text	

When a professional reviewer requires the input of a particular faculty member on a particular applicant, the document is moved to the "Referred to Special Admissions" state (Table 7.13). This state is given a one-month duration and both an overdue notification and follow-up to ensure a timely reply.

Table 7.13 The "Referred to Special Admissions" state.

Workflow name	WFUAdmissions
Form name	Application for Admission
State name	Referred to Special Admissions
State number	10
State owner	Faculty Reviewers

Table 7.13 The "Referred to Special Admissions" state. (continued)

State duration	1 Month
Send overdue notification?	Yes

State overdue notice text	Dear Faculty Member:
	    An application was forwarded to you approximately 1 month ago for your review and recommendation. To date, your input has not been recorded on the application. Please access this application at your earliest convenience. Thank you.
	    WFU Admissions Office

Send overdue follow-up?	Yes
Overdue follow-up duration	1 week

Overdue follow-up text	Dear Faculty Member:
	    An application was forwarded to you more than 1 month ago for your review and recommendation. To date, your input has not been recorded on the application. Please access this application immediately. Thank you.
	    WFU Admissions Office

Applicants that don't forward either their application fee or supporting information have their applications relegated to the "Inactive" state (Table 7.14). The experience of the admissions staff is that the majority of these applications are eventually reactivated. Thus, this state preserves the applicant information until it is either reactivated or deleted.

Table 7.14 The "Inactive" state.

Workflow name	WFUAdmissions
Form name	Application for Admission
State name	Inactive
State number	11
State owner	Admissions Professional Staff
State duration	12 Months
Send overdue notification?	No
State overdue notice text	
Send overdue follow-up?	No
Overdue follow-up duration	
Overdue follow-up text	

An application can be "Deleted" (Table 7.15) at any point during the workflow process. Once the document enters this state, a scheduled agent can physically delete the file.

Table 7.15 The "Deleted" state.

Workflow name	WFUAdmissions
Form name	Application for Admission
State name	Deleted
State number	12
State owner	Administrator
State duration	12 Months
Send overdue notification?	No
State overdue notice text	
Send overdue follow-up?	No
Overdue follow-up duration	
Overdue follow-up text	

That concludes the configuration of the state documents for the admissions application process. In the next section, we define the available actions.

7.2.2 Defining the Actions

The "Register" action (Table 7.16), similar to the first state, is not part of the workflow engine design. It is included mainly for documentation. This action is executed by the applicant during his/her first online editing session by clicking on the "Submit" button (not part of the workflow subform). The state change indicated is not explicit, but occurs because it is the default state of the workflow subform, and that subform is not visible when the document is initially composed. The notification setups for the state owner and document originator (i.e., the applicant) are misleading in this case, as the notifications are handled by the *registerApplicant* agent.

Table 7.16 The "Register" action.

Workflow name	WFUAdmissions
Form name	Application for Admission
Action number	001
Action name	Save
States associated with this action	Draft

Table 7.16 The "Register" action. (continued)

State change or Apply workflow logic rules	State Change
Changes to state name	Draft
Workflow logic rules	
Write to history log?	Yes
Processing options	
Notify state owner?	Yes
Notify first role member only?	No
State owner notification text	
Notify others via e-mail?	No
E-mail addresses to notify	
Group to notify	
Notify others text	
Notify originator?	Yes
Notify originator text	

The "Save As Draft" action (Table 7.17) is available to applicants so they may edit their applications multiple times before submission. There are no state changes, notifications, or history log entries associated with this action.

Table 7.17 The "Save As Draft" action.

Workflow name	WFUAdmissions
Form name	Application for Admission
Action number	002
Action name	Save As Draft
States associated with this action	Draft
State change or Apply workflow logic rules	State Change
Changes to state name	Draft
Workflow logic rules	
Write to history log?	No
Processing options	
Notify state owner?	No
Notify first role member only?	No
State owner notification text	
Notify others via e-mail?	No

Table 7.17 The "Save As Draft" action. (continued)

E-mail addresses to notify	
Group to notify	
Notify others text	
Notify originator?	No
Notify originator text	

The applicant selects the "Submit" action (Table 7.18) when they have completed their online application. This action: causes a state change to the document from "Draft" to "Submitted"; creates an entry in the history log; and notifies both the new document owner and the originator (i.e., the applicant).

Table 7.18 The "Submit" action.

Workflow name	WFUAdmissions
Form name	Application for Admission
Action number	003
Action name	Submit
States associated with this action	New Draft, Draft
State change or Apply workflow logic rules	State Change
Changes to state name	Submitted
Workflow logic rules	
Write to history log?	Yes
Processing options	
Notify state owner?	Yes
Notify first role member only?	
State owner notification text	A new application has been submitted for processing. Please access the WFU online site to access this application.
Notify others via e-mail?	No
E-mail addresses to notify	
Group to notify	
Notify others text	

Table 7.18 The "Submit" action. (continued)

Notify originator?	Yes
Notify originator text	Dear WFU Applicant: Thank you for your application for admission to WFU. We appreciate your interest and look forward to processing your application. Before your application can be processed the Admissions Office needs to receive: *Your transcripts, SAT scores, letters of recommendation, GPA, etc., from your High School Guidance Office. *A check or money order for $25 payable to WFU Admissions. These items must be received within the next 30 days. Again, thank you for your interest in WFU! WFU Admissions Office

The "Proceed" action (Table 7.19) is executed by the *AdmissionsSupportStaff* group after an application is submitted. As you can see from the original state diagram (Figure 6.7), there are two decision points that the document must pass through before it can enter either the "Under Evaluation" or "Inactive" states. "Proceed" is the valid action for the support staff when they record both the payment of the application fee and the reception of the student's information. These correspond to the two decision points in the diagram.

For this action, we specify workflow logic rules in place of an explicit state change. We then supply the XML that will be used by the rules engine to determine the next state for the document.

The XML rules check the status of the *ApplicationFee* and *InfoComplete* fields, setting the appropriate state number based on the field values. As the time period for the application fee to be paid is shorter than the time allowed for the student's information to arrive, that test is performed first in the code. When either the fee is paid or the information arrives, the support staff accesses the application to record the event. The staffer either marks the application fee as paid or the information received flag as complete. Then the staffer selects "Proceed". The rules engine fires the appropriate rule(s) and moves the document to either the "Waiting for Application Fee", "Waiting for Information", or "Under Evaluation" state.

Table 7.19 The "Proceed" action.

Workflow name	WFUAdmissions
Form name	Application for Admission
Action number	004
Action name	Proceed
States associated with this action	Waiting for Application Fee, Waiting for Information, Under Evaluation
State change or Apply workflow logic rules	Logic Rule
Changes to state name	

Workflow logic rules	`<?xml version="1.0" standalone="yes"?>`
	`<WORKFLOWLOGIC>`
	`<FACTS>`
	`    <FACT>`
	`        <NAME>ApplicationFee</NAME>`
	`        <VALUE>NotesField</VALUE>`
	`    </FACT>`
	`    <FACT>`
	`        <NAME>InfoComplete</NAME>`
	`        <VALUE>NotesField</VALUE>`
	`    </FACT>`
	`    <FACT>`
	`        <NAME>wfState_num</NAME>`
	`        <VALUE>NotesField</VALUE>`
	`    </FACT>`
	`</FACTS>`
	`<RULES>`
	`    <RULE TRANSITION="true">`
	`        <CONDITION>`
	`        <LHS>ApplicationFee</LHS>`
	`        <OP>EQUALTO</OP>`
	`        <RHS>1</RHS>`
	`        </CONDITION>`

Table 7.19 The "Proceed" action. (continued)

| | ```
      <ACTION>
        <LHS>wfState_num</LHS>
         <OP>EQUALTO</OP>
         <RHS>6</RHS>
      </ACTION>
   </RULE>
   <RULE TRANSITION="true">
      <CONDITION>
        <LHS>InfoComplete</LHS>
         <OP>EQUALTO</OP>
         <RHS>1</RHS>
      </CONDITION>
      <ACTION>
        <LHS>wfState_num</LHS>
         <OP>EQUALTO</OP>
         <RHS>7</RHS>
      </ACTION>
   </RULE>
   <RULE TRANSITION="true">
      <CONDITION>
        <LHS>InfoComplete</LHS>
         <OP>EQUALTO</OP>
         <RHS>2</RHS>
      </CONDITION>
      <ACTION>
        <LHS>wfState_num</LHS>
         <OP>EQUALTO</OP>
         <RHS>8</RHS>
      </ACTION>
   </RULE>
</RULES>
</WORKFLOWLOGIC>
``` |
| Write to history log? | No |
```

Table 7.19 The "Proceed" action. (continued)

Processing options	
Notify state owner?	No
Notify first role member only?	
State owner notification text	
Notify others via e-mail?	
E-mail addresses to notify	
Group to notify	
Notify others text	
Notify originator?	No
Notify originator text	

The "Accept" action (Table 7.20) is available to the professional staff after a document enters the "Under Evaluation" state.

Table 7.20 The "Accept" action.

Workflow name	WFUAdmissions
Form name	Application for Admission
Action number	005
Action name	Accept
States associated with this action	Submitted, Under Evaluation, Accepted
State change or Apply workflow logic rules	State Change
Changes to state name	Accepted
Workflow logic rules	
Write to history log?	Yes
Processing options	
Notify state owner?	No
Notify first role member only?	No
State owner notification text	
Notify others via e-mail?	No
E-mail addresses to notify	
Group to notify	

Table 7.20 The "Accept" action. (continued)

Notify others text	
Notify originator?	Yes

Notify originator text	Dear WFU Applicant:
	    Congratulations!
	    We are happy to inform you of your acceptance to Wyrcan Flowan University for the fall 2001 Semester. You will be receiving our WFU Acceptance Package in the mail within the next 2–4 weeks. This package should answer all of your questions about what to do next along with information on WFU housing, financial aid, and a student life brochure. In the interim, please feel call us with any questions or visit out WFU web site (www.wyrcanflowan.edu) to post a question in our new student discussion forum.
	    We look forward to hearing from you.
	    WFU Admissions Office

The "Reject" action (Table 7.21) marks another termination point. This action writes an entry in the history file and then generates a "Dear Applicant" letter.

Table 7.21 The "Reject" action.

Workflow name	WFUAdmissions
Form name	Application for Admission
Action number	006
Action name	Reject
States associated with this action	Rejected
State change or Apply workflow logic rules	State Change
Changes to state name	Rejected
Workflow logic rules	
Write to history log?	Yes
Processing options	
Notify state owner?	No
Notify first role member only?	No
State owner notification text	
Notify others via e-mail?	No

Table 7.21 The "Reject" action. (continued)

E-mail addresses to notify	
Group to notify	
Notify others text	
Notify originator?	Yes

Notify originator text	Dear Applicant:
	    We regret to inform you that your application was not accepted for admission to Wyrcan Flowan University. We thank you for your interest in WFU and wish you the best in your future educational pursuits.
	    WFU Admissions Office

The "Refer to Special Admissions" action (Table 7.22) is available to the professional staff when they want to solicit feedback on a particular application from a faculty member. For simplicity, we are representing faculty as a single role. In reality, this would probably be defined as multiple groups or individuals, organized by department or discipline.

Table 7.22 The "Refer to Special Admissions" action.

Workflow name	WFUAdmissions
Form name	Application for Admission
Action number	007
Action name	Refer to Special Admissions
States associated with this action	Under Evaluation, Referred to Special Admissions
State change or Apply workflow logic rules	State Change
Changes to state name	Referred to Special Admissions
Workflow logic rules	
Write to history log?	Yes
Processing options	
Notify state owner?	Yes
Notify first role member only?	No

Table 7.22 The "Refer to Special Admissions" action. (continued)

State owner notification text	Dear Faculty Member: An application for admission has been assigned to you for your input. Please access the WFU online admissions site to review the application. You have 30 days in which to review the application and complete your comments. Thank you. Admissions Office
Notify others via e-mail?	No
E-mail addresses to notify	
Group to notify	
Notify others text	
Notify originator?	No
Notify originator text	

The "Defer" action (Table 7.23) is used by professional staffers to postpone final resolution on an application. An outside event or influence (summer school, low volume of applications, etc.) may justify reactivating a borderline acceptable application.

Table 7.23 The "Defer" action.

Workflow name	WFUAdmissions
Form name	Application for Admission
Action number	008
Action name	Defer
States associated with this action	Under Evaluation, Deferred
State change or Apply workflow logic rules	State Change
Changes to state name	Deferred
Workflow logic rules	
Write to history log?	Yes
Processing options	
Notify state owner?	No
Notify first role member only?	No
State owner notification text	
Notify others via e-mail?	No

Table 7.23 The "Defer" action. (continued)

E-mail addresses to notify	
Group to notify	
Notify others text	
Notify originator?	No
Notify originator text	

Our last action, "Delete" (Table 7.24), frequently *is* the last action. This action, available only to Administrators, changes the state of the document. This removes it from most of the views. The document itself is not deleted, as the school must retain all applications for five years.

Table 7.24 The "Delete" action.

Form name	Application for Admission
Action number	999
Action name	Delete
States associated with this action	Under Evaluation, Deleted
State change or Apply workflow logic rules	State Change
Changes to state name	Deleted
Workflow logic rules	
Write to history log?	Yes
Processing options	
Notify state owner?	No
Notify first role member only?	No
State owner notification text	
Notify others via e-mail?	No
E-mail addresses to notify	
Group to notify	
Notify others text	
Notify originator?	No
Notify originator text	

7.2.3 Specifying the Roles

The roles in our application are the same roles depicted in the swimlane activity diagram of Chapter 6 (Figure 6.6).

- Applicant

- Admissions Support Staff
- Information Providers
- Admissions Professional Staff
- Faculty

As the online admissions system will not be directly interacting with the any information providers, we will not create a role document for this group. We will, however, create an additional role named Administrator. This role will add workflow system administration capability and security to our application. Administrators have the ability to execute any action on any document.

Let's review the completed role documents for this workflow configuration, starting with the new Administrator role. The definition of a role is similar to the configuration of actions and states. That is, we identify the workflow in which members of this role will be actors, along with a role number and the form name of the workflow-enabled document.

Note that we check off all actions to allow Administrators the ability to handle a document in any state (Table 7.25).

The next field lets us use the role document itself to define the members of a role, rather than using the Domino directory (which often requires the services of your network Notes administrators). This feature is well-suited for small departmental workflows, or in situations in which role members change frequently and the request to update the group in the Domino directory may not be turned around quickly.

Table 7.25 The "Administrator" role.

Workflow name	WFUAdmissions
Form name	Application for Admission
Role name	Administrator
Role number	001
Actions associated with this role	Accept, Defer, Delete, Proceed, Reject, Save, Save As Draft, Submit, Refer to Special Admissions
Role members defined	Role Document
Role NAB group	
Role members	Dick Lam, Dan Giblin

As you may recall from Chapter 6, the Admissions Professional Staff consists of the Director of Admissions, four assistant directors, and five associate directors. These groups are already defined in the Domino directory, so we simply specify the existing group names in the role definition document (Table 7.26).

Table 7.26 The "Admissions Professional Staff" role.

Workflow name	WFUAdmissions
Form name	Application for Admission
Role name	AdmissionsProfessionalStaff
Role number	002
Actions associated with this role	Accept, Defer, Reject, Refer to Special Admissions
Role members defined	Domino Directory
Role NAB group	AdmissionsProfessionalStaff
Role members	

The Director of Admissions also has a support staff that is responsible for preparing the applications for evaluation by the professional staff. Support staffers review the applications for completeness, record the reception of the application fee, and enter the student's transcripts, SAT and other test scores, etc. Because this group may be more dynamic in membership than the professional staff, the support staff members are defined directly within the role document (Table 7.27).

Table 7.27 The "Admissions Support Staff" role.

Workflow name	WFUAdmissions
Form name	Application for Admission
Role name	AdmissionsSupportStaff
Role number	003
Actions associated with this role	Proceed
Role members defined	Document
Role NAB group	
Role members	Jim Samuels Caitlin Fiona Pete Carney Randi Perez Norm Stahlheber

The Faculty Reviewers role consists of faculty members who occasionally receive applications for review from the admissions office. Faculty members are usually organized by discipline or department in existing Domino directory groups (Table 7.28).

Table 7.28 The "Faculty Reviewers" role.

Workflow name	WFUAdmissions
Form name	Application for Admission
Role name	FacultyReviewers
Role number	004
Actions associated with this role	Save
Role members defined	Domino Directory
Role NAB group	FacultyApplicationReviewers
Role members	

Our last, but certainly not least, role is that of the applicant. For this role, we indicate that the role is defined in the role document and use the wildcard character standard of *.* to indicate that any user of the system can function in this role. This includes unregistered users, as well (Table 7.29).

Table 7.29 The "Applicant" role.

Workflow name	WFUAdmissions
Form name	Application for Admission
Role name	Applicant
Role number	005
Actions associated with this role	Save As Draft, Submit
Role members defined	Document
Role NAB group	
Role members	*.*

That completes our configuration of the profile, actions, states, and roles that define the workflow-enabled online admissions process.

7.3 Agents

Our case study application uses five different agents. The first agent (*registerApplicant*) was presented in Chapter 6. That agent continues to function in the *WebQuerySave* event of the "Application for Admission" form during its first editing session. The next three agents were first presented in Chapter 5: *wfCurActionsList*, *wfProcess*, and *wfProcessForm*.

As explained earlier in this chapter, when a user initially visits the site and creates an online application, the first and only processing option available is to press the "Submit" button. This button does not actually submit the application for review, but saves the document and triggers the *registerApplicant* agent.

On all subsequent processing sessions, the *WebQuerySave* agent *wfProcess* executes to handle all of the workflow processing of the document. This task is really split between two agents: the Java agent named *wfProcess*, and a LotusScript agent named *wfProcessForm*. The Java agent handles the explicit state changes or rule processing defined for the action being performed. It then calls *wfProcessForm*, passing it the note ID of the current document. This agent finds the application document based on the note ID and completes any other processing tasks, such as notifications and history log entries.

7.3.1 The *wfOverDueMonitor* Agent

One of the key considerations when designing a workflow process is to ensure that the documents do indeed flow through the process. As part of the state descriptions, we identified a duration period for each state, along with a text message to be sent to the state owner if the duration period is exceeded.

In the event that this "duration exceeded" notice fails to generate a response from the state owner, a follow-up duration and text message are specified. This enables the framework to send multiple follow-up notices until the document owner performs some action. Note that this type of notification should be used sparingly so it does not lose its effectiveness or alienate your users.

A new agent (*wfOverDueMonitor*) has been created to monitor the workflow-enabled documents in the application. This agent is scheduled to run on a nightly basis, generating the appropriate notifications when necessary. Listing 7.1 shows the code for this agent (written in LotusScript).

Listing 7.1 The *wfOverDueMonitor* agent.

```
'****************************************************************************
'
' Module: wfOverDueMonitor.ls
' Author: Dan Giblin
'
```

Listing 7.1 The *wfOverDueMonitor* agent. (continued)

```
' Description:   This agent creates a collection of all documents whose next
'                action date is older than today. It then iterates through the
'                collection to see if the notification switch is turned on for
'                that particular state, or if an overdue notification was
'                already sent. If a notification was already sent, it checks
'                to see if the follow-up option is turned on for that state,
'                and if the follow-up time period has been exceeded yet.
'                Follow-up notifications are sent according to the schedule
'                in the state document.
'
'*******************************************************************************


Sub Initialize
    Dim s As NotesSession, db As NotesDatabase, note As NotesDocument
    Dim wf As WorkflowDocument, sUser As String, sNewOwner As String
    Dim sOriginator As String, notifier As Notification, key As String
    On Error Goto wfOverDueMonitorError

     ' get the current session parameters
    Set s = New NotesSession
    Set db = s.CurrentDatabase
    Set documents = db.Search("@Today => @Date(wfNextActionDate)", Nothing, 0)

     ' create the Notification instance
    key = ""
    sNewOwner = ""
    sOriginator = ""
    Set notifier = New Notification(db, key, sNewOwner, sOriginator)

     ' make sure there are documents to process
    If documents.Count = 0 Then
        Exit Sub
    End If
```

Listing 7.1 The *wfOverDueMonitor* **agent. (continued)**

```
' iterate through the document collection
For d = 1 To documents.Count
    Set note = documents.GetNthDocument(d)
    Set wf = New WorkflowDocument(s,db,note)


    ' is the send notice turned on for this state?
    If wf.getSendOverDueNotice = 1 Then
        Dim vtemp As Variant

        vtemp = wf.getLateNoticeSentDate
        If vtemp(0) = "" Then
            ' notice has not already been sent - create a notification
            sActors = wf.getNewOwner
            sSubject = "Overdue Application Notification"
            sText = wf.getOverDueText
            Call notifier.CreateNotification(sSubject, sText, sActors)

            ' update document with today's date in the
            ' late notice sent field
            Call wf.setLateNoticeSentDate
            Call note.Save(True,True)
        Elseif wf.getSendOverDueFollowup = 1 Then
            'find out if the send followup flag is on
            Dim q As Integer
            Dim x As Integer
            Dim v As Variant

            ' what is the duration for the 2nd late notice?
            q = wf.calcFollowupDurationinDays
            v = wf.getLateNoticeSentDate
            x = (Today -  v(0))
```

Listing 7.1 The *wfOverDueMonitor* agent. (continued)

```
            If x >= q Then
                    ' send notice
                sActors = wf.getNewOwner
                sSubject = "Overdue Application Notification"
                sText = wf.getFollowupText
                Call notifier.CreateNotification(sSubject, sText, _
                sActors)
                Call wf.setLateNoticeSentDate
                Call note.Save(True,True)
            End If
        End If
    End If
    Next


wfOverDueMonitorError:
    Exit Sub
End Sub
```

wfOverDueMonitor starts by creating a collection of documents whose next action date (*wfNextActionDate*) is less than or equal to the current date. If no documents meet this criterion, the agent terminates. For those documents having a next action date of today or earlier, the agent first checks to see if a late notice has already been sent. That check is accomplished by examining the *wfLateNoticeSentDate* field. If this field is blank, a notice has not been sent for the document, and one is constructed. We get the owner of the document based on the current state and also the notification text contained in this state's configuration document. These are passed to the notification database, along with a subject title. The notification database handles the rest of the work in generating and sending the e-mail message. The agent finishes its task by updating the *wfLateNoticeSentDate* with the current date and saving the document.

While iterating through the document collection, if we find a document that has already had a late notice sent, we check the state document to see if the *wfs_SendOverDueFollowup* notification flag is set. If so, we check the duration period for this follow-up and calculate the number of days before the next notice. We also calculate the difference in days between the current date and the date in the *wfLateNoticeSentDate* field. If the time for a new notice has been reached, we gather the state owner, overdue message text, and a subject title to pass to the notification engine for processing. The agent then finishes processing, as in the previously discussed case.

This agent should be run on a schedule appropriate for the particular workflow process. For this admissions application, it runs nightly, based on the volume of documents expected and the need to provide a timely turn-around for each document submission.

7.4 A Test Drive

With the workflow configuration completed and the agents in place, we can take the admissions site for a test drive. The test drive will create an "Application for Admission" document and monitor it as it proceeds through to the "Submitted" state.

Starting at the Wyrcan Flowan University Admissions Office home page, we start the application process as an unauthenticated user (Figure 7.3).

Figure 7.3 The home page for the WFU Admissions office.

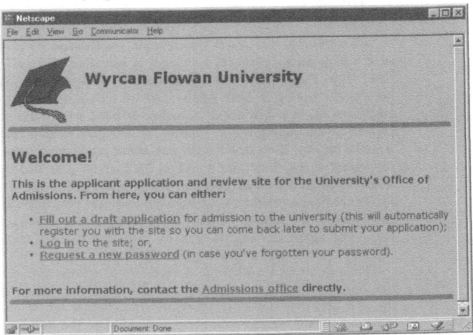

Clicking on the "Fill out a draft application" link brings up the online application form.

For the user's first editing session, an expanded help section appears at the top of the application form (Figure 7.4). The input fields remain the same as the application form in the Chapter 6 database. After completing the application, the user clicks the "Submit" button at the bottom of the form.

Figure 7.4 A blank application for admission form.

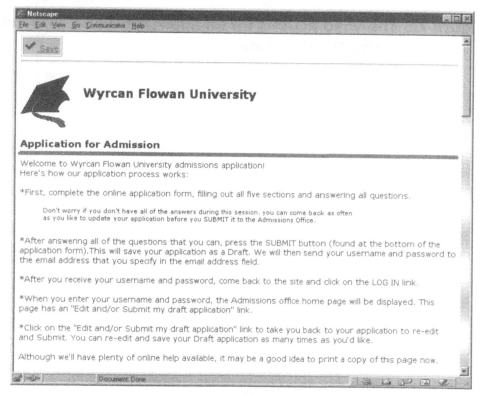

This initial submission of the form launches the *registerApplicant* agent. The applicant is granted access to the site as a user, and receives e-mail containing his/her user name and password.

On the student's return visit to the site, he/she logs in with their new user name and password to complete and submit their application. The workflow engine is active at this point, so the action selection radio buttons on the workflow subform are displayed to the user (Figure 7.5). The valid actions are computed by the *wfGetCurActionsList* agent in the *WebQueryOpen* event of the form. For our workflow configuration, the valid actions for an applicant editing an application in the "Draft" state are "Save As Draft" and "Submit".

After selecting an action and clicking the "Proceed" button, the document is processed by the workflow engine according to the parameters set up in our action and state documents. In this example, the user selects the "Submit" action, causing a state change from "Draft" to "Submitted". The admissions support staff is notified, and the applicant receives an e-mail acknowledgement.

Figure 7.5 The valid actions for an applicant when a document is in the "Draft" state.

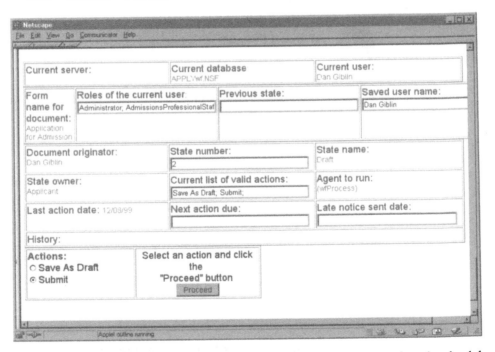

At any time, the status of the documents in process can be checked by returning to the home page and clicking on the "Review applications for admission" link. Selecting this link displays the *wfFrameset* page, containing the "Applications" view in the right-hand frame. We can see from the status column that the document was processed and moved into the "Submitted" state (Figure 7.6).

Figure 7.6 The document status view.

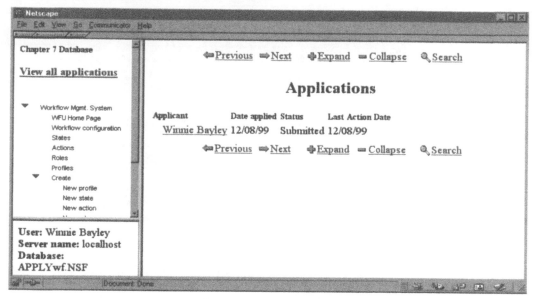

If we log back into the site as an Administrator, we can take a closer look at what has happened to the document at this point. Open the application document and scroll down to the visible *wf_SubForm* to view the present history of the document and examine the computed values for the other fields (shown in Figure 7.7).

Figure 7.7 The workflow subform (unhidden) in an active workflow document.

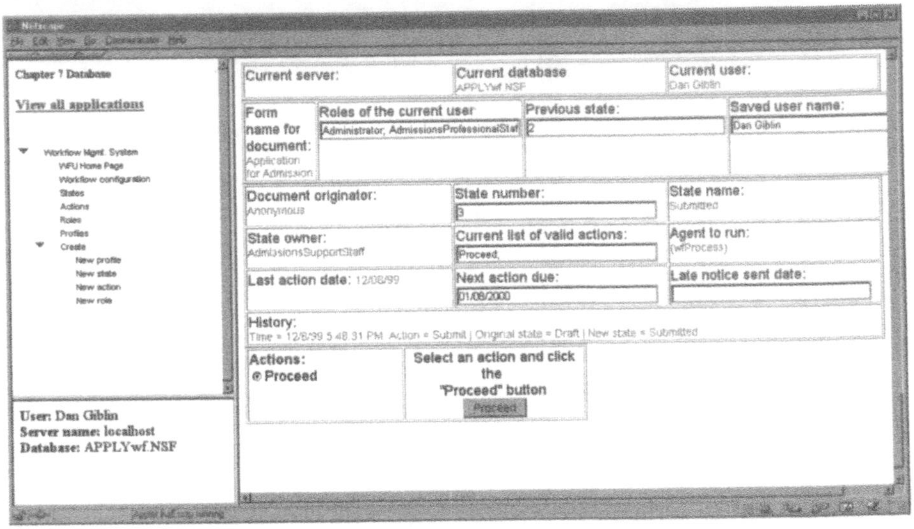

The *wfRole* field is populated with the text list "Administrator; AdmissionsProfessionalStaff; Applicant," as those are all of the valid roles for this authenticated user.

Note that the owner of the document in this state is the AdmissionsSupportStaff group, and the only valid action is "Proceed". The date of the last action has been recorded and the next action date has been calculated. Finally, the *wfHistory* field shows a summary of the workflow-related events.

7.5 Summary

This concludes our review of the workflow-enabled online admissions system for our case study. It also concludes this book. We hope that we have met our goals to introduce the basic concepts of workflow as it relates to business processes, and the power of Domino as a web-based workflow application development and hosting environment. We also hope we have adequately illustrated and explained the development of a practical workflow management framework that you can extend and use in your own Domino applications. We wish you the best success in applying these workflow concepts and the framework to your next web-based workflow application.

Appendix A

Workflow Management Systems

In this appendix, we present a brief introduction to some of the workflow management systems on the market, and organizations concerned with workflow processes and standardization. The products discussed are as follows.

- Domino Workflow
- Percussion PowerFlow

  Overviews are presented for the following workflow organizations.
- SWAP — Simple Workflow Access Protocol
- WARIA — Workflow and Reengineering International Association
- WfMC — Workflow Management Coalition

A.1 Domino Workflow 2.0

In April of 1999, Lotus Corporation announced the acquisition of ONEStone Holding, Inc. ONEStone's primary product was Prozessware, a workflow management system designed for Notes/Domino. This product was renamed Domino Workflow[1].

Some of Domino Workflow's features include the following.

- Graphical process design and management
- Automated document routing
- Binder management
- Surrogate management
- Time management
- Automated access
- Job monitoring
- Ad hoc exception handling
- Distributed processes support

Domino Workflow consists of the following three components.

1. Domino Workflow Architect The Architect provides an intuitive, graphical "drag-n-drop" interface that enables system designers to quickly design and manage workflow processes. Designers can specify workflow process logic, participants, and the rules governing participation without the need for programming. By representing the entire workflow process in one window view, developers can modify all aspects of the workflow process easily. The Architect then automatically activates process definitions for the Domino Workflow Engine, and offers a process debugging feature to assure that process logic is complete and error-free prior to activation.

2. Domino Workflow Engine The Engine includes five Domino databases that implement and store the process logic, organization information, process-related information objects, workflow instances, audit trails, and archives. Developers save and reuse process logic and components utilizing Notes/Domino object services and design tools, including Domino Designer. No training is required, as the engine uses familiar Notes/Domino administration and management services. The core of the workflow system, the Engine, is comprised of three Domino applications.

The **process definition database** stores the process steps, their sequences, and routing rules;

The **organization directory database** allows the definition, grouping and assignment of workflow participants and resources.

The **application template** contains predefined application building blocks, and stores or links to information objects that interact with participants in performing workflow activities.

3. Domino Workflow View The View allows a participant to view the status and context of activities within a workflow process.

A.2 Percussion PowerFlow

Percussion's PowerFlow[2] uses a visual drag-and-drop designer for workflow enabling an application. It works with any application design and allows you to change and enhance the automated processes of currently deployed applications. Its active state engine runs on a Domino server and uses Percussion's Event Management technology to monitor process-enabled applications for document events. In addition, rules may be defined in the PowerFlow Manager database.

Key features include the following.

- Any process may be automated "as is", without spending time reengineering or using special business process design elements.
- Integrated user management enables automatic synchronization of any Person Management database with the PowerFlow hierarchy.
- Workflow scripting is not necessary.
- Reusable objects can be used to branch documents to a logical set of states and transitions that will return the document to the main workflow.
- Shared Notes databases contain document-based workflow.
- Automatic notification through customized e-mail notifies participants of documents awaiting their review.
- Audit trails visually identify the flow of each document, including who acted, when, and under what conditions.
- Charts show performance times by person, state, and for the workflow as a whole.
- People assigned to roles in the active process may be changed, and the engine automatically reassigns the work, even for documents already in process.
- Participants in a workflow process can delegate their work.
- Rules and routes can change without changing the end-user interface.

A.3 SWAP — Simple Workflow Access Protocol

The main objective of the SWAP[3] working group is to define requirements for developing an Internet-based workflow access protocol. The protocol is designed to instantiate, control, and monitor workflow process instances across heterogeneous workflow engines.

There is no standard way to communicate and interoperate across heterogeneous workflow engines to coordinate and execute work process instances. Most of the solutions existing today are vendor-specific. Thus, workflow vendors need to support multiple protocols to enable interoperability among workflow systems. SWAP is developing the goals and requirements for workflow systems interoperability. A further goal is to define an application-level workflow interoperability protocol.

A.4 WARIA — Workflow and Reengineering International Association

The charter of the Workflow and Reengineering International Association[4] is to identify and clarify issues that are common to all users of workflow, electronic commerce and those who are in the process of reengineering their organizations. The association facilitates opportunities for members to discuss and share their experiences.

WARIA's mission is to make sense of the technologies and ideas around the intersection of business process reengineering, workflow, knowledge management, and electronic commerce. The goal is to clarify issues through experience-sharing, product evaluations, networking between users and vendors, education, and training.

WARIA produces a series of books about workflow and knowledge management[5–7].

A.5 WfMC — Workflow Management Coalition

The Workflow Management Coalition[8], founded in August 1993, is a nonprofit, international organization of workflow vendors, users, analysts, and university and research groups. The Coalition's mission is to promote and develop the use of workflow through the establishment of standards for software terminology, interoperability, and connectivity among workflow products. There are over one hundred and thirty worldwide members. The

WfMC has established itself as the primary standards body for this rapidly expanding software market.

A.6 References

1 Lotus Development Corporation. Domino Workflow. [http://www.lotus.com/home.nsf/welcome/domworkflow]

2 Percussion Software. [http://www.percussion.com]

3 SWAP. [http://www.ics.uci.edu/~ietfswap/]

4 WARIA. [http://www.waria.com]

5 Layna Fischer. 1997. *Innovation and Excellence in Workflow and Imaging*, Volume I. WARIA.

6 Layna Fischer. 1998. *Innovation and Excellence in Workflow and Imaging*, Volume II. WARIA.

7 Layna Fischer. 1999. *Innovation and Excellence in Workflow/Process Management and Document/Knowledge Management*, Volume III. WARIA.

8 WfMC. [http://www.aiim.org/wfmc/]

Index

A

X

What's on the CD-ROM?

All of the code in this book is included on the accompanying CD-ROM. The code examples are organized by chapter, with each chapter having its own corresponding subdirectory. The table below shows the naming conventions followed for the files on the CD-ROM.

CD-ROM file naming conventions.

File Type	Naming Convention
LotusScript agent source file	*name*.ls (**e.g.,** test.ls)
Java agent source file	*name*.java (**e.g.,** test.java)
Java agent class file	*name*.class (**e.g.,** test.class)
JAR file (Java archive)	*name*.jar (**e.g.,** test.jar)
Notes template	*name*.ntf (**e.g.,** notify.ntf)
Notes database	*name*.nsf (**e.g.,** notify.nsf)

The workflow management tools, templates, and code discussed in this book are based on the Lotus Notes/Domino environment. All of the Notes/Domino examples were created on the version of Notes known as R5, using the Domino Designer application. The Web-based example programs were tested on either Windows 95 or Windows 98 client machines, running either Netscape Navigator version 4.6+ or Internet Explorer version 5.0+. The client machines were tested against server applications running on Windows NT Server version 4.0 with service pack 5, running Domino version 5.0.1 or greater.

For more information on the environments, languages, and code, see page 14 of Chapter 1. For more information on the CD-ROM contents and how to set up the example databases, see the helpful readme.doc on the accompanying CD-ROM.

Printed and bound by CPI Group (UK) Ltd, Croydon, CR0 4YY

21/10/2024

01777098-0002